Encyclopedia of Electronic Components Volume 3

Charles Platt and Fredrik Jansson

MAKER MEDIA™
SAN FRANCISCO, CA

Encyclopedia of Electronic Components, Volume 3

by Charles Platt

Printed in Canada.

Published by Maker Media, Inc., 1160 Battery Street East, Suite 125, San Francisco, CA 94111.

Maker Media books may be purchased for educational, business, or sales promotional use. Online editions are also available for most titles (*http://safaribooksonline.com*). For more information, contact O'Reilly Media's institutional sales department: 800-998-9938 or *corporate@oreilly.com*.

Editor: Brian Jepson
Production Editor: Melanie Yarbrough
Copyeditor: Christina Edwards
Proofreader: Charles Roumeliotis

Indexer: Charles Platt
Interior Designer: David Futato
Cover Designer: Karen Montgomery
Illustrator: Charles Platt

April 2016:   First Edition

Revision History for the First Edition

2016-04-05:   First Release

See *http://oreilly.com/catalog/errata.csp?isbn=9781449334314* for release details.

978-1-449-33431-4

[TI]

To Brian Jepson

Table of Contents

Preface

This third and final volume of the *Encyclopedia of Electronic Components* is devoted entirely to sensors.

Two factors have caused very significant changes in the field of sensors since the 1980s. First, features such as antilock braking, airbags, and emissions controls stimulated the development of low-priced sensors for automotive applications. Many of these sensors were fabricated in silicon as MEMS (microelectromechanical) devices.

The second wave began in 2007 when MEMS sensors were installed in the iPhone. A modern phone may contain almost a dozen different types of sensors, and their size and price have been driven down to a point that would have been unimaginable 20 years previously.

Many MEMS sensors are now as cheap as basic semiconductor components such as a voltage regulator or a logic chip, and they are easy to use in conjunction with microcontrollers. In this Encyclopedia, we have allocated significant space to this segment of the market, hoping that the specific products that we have chosen will remain popular and available for at least the next decade.

In addition, we have devoted space to older components where durability has been proven.

Purpose

While much of the information in this volume can be found dispersed among datasheets, introductory texts, Internet sites, and technical briefings published by manufacturers, we believe there is a real need for a durable resource that assembles all the relevant data in one place, properly organized and verified, including details that may be hard to find elsewhere.

This volume may also serve a useful purpose by attempting to categorize and classify components in a field that is remarkably chaotic. For example, is an *object presence sensor* different from a *proximity sensor*? Some manufacturers seem to think so; others disagree. Understanding the distinctions and the underlying principles can be important if you are trying to decide which sensor to use.

Sensor terminology can also be confusing. To take another example, what is the difference between a *reflective interrupter*, a *reflective object sensor*, a *reflective optical sensor*, a *reflective photointerrupter*, and an *opt-pass sensor*? These terms are used in various datasheets to describe components that are all *retroreflective sensors*. Understanding the proliferating variety of terminology can be essential if you simply want to find something in a product index.

Organization

As in volumes 1 and 2, this volume is organized by subject. For example, if you want to measure temperature, you'll find the entries for a thermistor and a thermocouple next to each other, in an entire section devoted to the sensing of heat. This will help you to compare capabilities and choose the component that best suits your application.

The subject path leading to each sensor is shown at the top of the first page of each entry. For gas flow rate, for instance, you would follow this path:

```
fluid > gas > flow rate
```

Note that the word "fluid" is properly used to include gases as well as liquids.

Exceptions and Conflicts

Unfortunately, some sensors are not easily categorized. There are four problems in this area.

1. What Does a Sensor Really Sense?

A GPS chip is a radio receiver, picking up transmissions from satellites. Does this mean it should be categorized as a sensor of radio waves? No, its purpose is to tell you your location. Therefore, it is categorized as a location sensor. This leads to the first general rule: sensors are categorized by their primary purpose. Secondary purposes may be found in the index.

2. How Many Sensors Are in a Sensor?

Many surface-mount chips perform more than one sensing function. For example, an inertial measurement unit (often identified by its acronym, *IMU*) can contain three gyroscope sensors and three accelerometers—and may contain three magnetometers, too. How should it be categorized?

The answer is that an IMU will be mentioned in more than one entry in the Encyclopedia, because it performs more than one function; but it will not have its own separate entry, because each entry in the Encyclopedia is for a single primary sensing function.

The names of multisensor chips are, of course, included in the index.

3. How Many Stimuli Can One Sensor Sense?

A single sensing element may be used in multiple different types of sensors. The most notable example is the Hall-effect sensor, which can be found in magnetometers, object presence sensors, speed sensors, current sensors, and dozens more. Modern automobiles can contain Hall-effect sensors everywhere from the ignition system to the trunk-locking mechanism. If you are using a hard drive with rotating platters, it probably contains a Hall-effect sensor to monitor the speed of rotation. If you have a generic computer keyboard, each keypress is probably detected with a Hall-effect sensor.

Bearing this in mind, how should a Hall-effect sensor be classified? And where should you expect to find an explanation of how it works?

The answer is that where different types of components contain the same type of sensing element, the entry for each component will include a cross-reference to one location where the sensing element is explained in detail.

This location will be chosen for its relevance. Thus, Hall-effect sensors are explained in the entry for **object presence** sensors, because this is their primary function. While it is true that a Hall-effect sensor works by detecting a magnetic field, that is not its most common application.

4. Too Many Sensors!

Wikipedia lists more than 100 general types of sensors (*http://bit.ly/1WJ9P12*), and even that list is probably not complete. Consequently, we had to pick and choose. Some of the decisions may seem arbitrary, but all of them were made on the grounds of practicality. There were three principles for deciding what to include and what to leave out.

1. Is it a component? We are more interested in board-mounted components than in packaged products that happen to contain sensors. For instance, a thermocouple is often enclosed in a tubular steel probe, and its wire is often plugged in to a specially designed meter that displays temperature. While we do include a photograph of a probe, we are primarily interested in the welded wires of the thermocouple inside it.

2. How much does it cost? An industrial ultrasonic sensor to check items on a factory conveyor belt will be sealed into a module with a waterproof grommet around a shielded cable—which is all very nice, but will not be very affordable. This Encyclopedia is more interested in board-mountable components for one-tenth of the price.

3. How many people are likely to want it? The stock of each type of sensor was checked on component vendor sites. If a sensor wasn't in the inventory, or if only a couple of variants were stocked, we concluded that the limited demand probably didn't justify including it here. For example, a Ferraris acceleration sensor responds to eddy currents in a rotating motor shaft, as a way of measuring vibration in the shaft. This is a really interesting device, but is unlikely to be on most people's shopping lists.

Volume Contents

Having explained the organization of this book and our decisions to include or omit various components, we now present a summary of the contents of all three Encyclopedia volumes:

Volume 1

Power; electromagnetic devices; discrete semiconductors.

The *power* category includes sources of electricity and methods to distribute, store, interrupt, convert, and regulate power. The *electromagnetism* category includes devices that exert force linearly, and others that create a turning force. *Discrete semiconductors* include the primary types of diodes and transistors. See Figure P-1 for a contents listing.

Primary Category	Secondary Category	Component Type
power	source	battery
	connection	jumper
		fuse
		pushbutton
		switch
		rotary switch
		rotational encoder
	moderation	relay
		resistor
		potentiometer
		capacitor
		variable capacitor
	conversion	inductor
		AC-AC transformer
		AC-DC power supply
		DC-DC converter
		DC-AC inverter
	regulation	voltage regulator
electro-magnetism	linear output	electromagnet
		solenoid
	rotational output	DC motor
		AC motor
		servo motor
		stepper motor
discrete semi-conductor	single junction	diode
		unijunction transistor
	multi-junction	bipolar transistor
		field-effect transistor

Figure P-1 *The subject-oriented organization of categories and entries in Volume 1 of this Encyclopedia.*

Volume 2

Thyristors (SCRs, diacs, and triacs); integrated circuits; light sources, indicators, and displays; and sound sources.

Integrated circuits are divided into analog and digital components. *Light sources, indicators, and displays* are divided into reflective displays, single sources of light, and displays that emit light. *Sound sources* are divided into those that create sound, and those that reproduce sound. A contents listing for Volume 2 appears in Figure P-2.

Volume 3

All the most common types of sensing devices, including those that detect location, presence, proximity, orientation, oscillation, force, load, human input, liquid properties, gas types and concentrations, pressure, flow rate, light, heat, sound, and electricity. A contents listing for Volume 3 appears in Figure P-3.

Method

Reference Versus Tutorial

As its title suggests, this is a reference book, not a tutorial. A tutorial such as *Make: Electronics* begins with elementary concepts and builds sequentially toward concepts that are more advanced. A reference book assumes that you may dip into the text at any point, learn what you need to know, and then put the book aside. If you choose to read it straight through from beginning to end, you will find some repetition, as each entry is intended to be self-sufficient, requiring minimal reference to other entries.

Theory and Practice

This book is oriented toward practicality rather than theory. We assume that the reader mostly wants to know how to use electronic components, rather than why they work the way they do. Consequently we do not include detailed proofs of formulae or definitions rooted in electrical theory.

Primary Category	Secondary Category	Component Type
discrete semi-conductor	thyristor	SCR
		diac
		triac
integrated circuit	analog	solid-state relay
		optocoupler
		comparator
		op-amp
		digital potentiometer
		timer
	digital	logic gate
		flip-flop
		shift register
		counter
		encoder
		decoder
		multiplexer
light source, indicator or display	reflective	LCD
	single source	incandescent lamp
		neon bulb
		fluorescent light
		laser
		LED indicator
		LED area lighting
	multi-source or panel	LED display
		vacuum-fluorescent
		electroluminescence
sound source	audio alert	transducer
		audio indicator
	reproducer	headphone
		speaker

Figure P-2 *The subject-oriented organization of categories and entries in Volume 2.*

Primary Category	Attribute to be Sensed	Type of Sensor
spatial	location	GPS
		magnetometer
	presence	object presence
		passive infrared
	distance	object proximity
		linear position
	orientation	rotary position
		tilt
		gyroscope
		accelerometer
mechanical	oscillation	vibration
	force	force
	human input	single touch
		touch screen
fluid	liquid	liquid level
		liquid flow rate
	gas/liquid	pressure
	gas	gas concentration
		gas flow rate
radiation	light	photoresistor
		photodiode
		phototransistor
	heat	NTC thermistor
		PTC thermistor
		thermocouple
		RTD
		semiconductor
		infrared temperature
	sound	microphone
electricity	metering	current
		voltage

Figure P-3 *The subject-oriented organization of categories and entries in Volume 3.*

Sensor Output

In Volumes 1 and 2 of the *Encyclopedia*, each entry included hints on how to use a component. However, many sensors have identical forms of output, which are processed in a similar way. To avoid repetition, general guidance for using nine principal types of sensor outputs has been placed in Appendix A at the back of this volume.

For example, many sensors provide an analog voltage output that varies with the phenomenon that is being sensed. In Appendix A, you will find suggestions on how to adjust the range of the output, if necessary, or how to digitize it with an analog-to-digital converter.

You will also find a comparison between serial protocols such as I2C and SPI, both of which are commonly used when a microcontroller communicates with a digital sensor via a bus.

Glossary

In the world of sensors, many terms tend to recur. *Hysteresis* is one; *MEMS* is another. Rather than define these terms repeatedly, some quick definitions are gathered in a Glossary. Please remember the existence of the glossary if you encounter a term that is unfamiliar. See Glossary.

In many instances, terms that are italicized in the text are defined in the glossary.

Typographical Conventions

Within each entry, **bold type** is used for the first occurrence in each entry of the name of a component that has its own entry elsewhere. Other important electronics terms or component names may be presented in *italics*.

The names of components, and the categories to which they belong, are all set in lowercase type, except where a term is normally capitalized because it is an acronym or a trademark, or contains a proper noun. The term *Hall effect*, for instance, has an initial cap because it is named after a person named Hall. The term *GPS* is all in caps, because it is an acronym; but *psi* (meaning pounds per square inch) remains in lowercase, because even though it is an acronym, the lowercase form is more common.

The situation is different when specifying units that are named after electrical pioneers. All of these units should be lowercased when spelled out. Thus, when referring to the SI unit of force, it is "the newton." However, where a unit named after a person is abbreviated, the abbreviation is capitalized, as in N for newtons, Hz for hertz, Pa for pascals, and A for amperes.

Mathematical Syntax

In mathematical formulae, we have used the style that is common in programming languages. The * (asterisk) is used as a multiplication symbol, while the / (forward slash) is used as a division symbol. Where some terms are in parentheses, they must be dealt with first. Where parentheses are inside parentheses, the innermost ones must be dealt with first. Consider this example:

```
A = 30 / (7 + (4 * 2) )
```

You would begin by multiplying 4 times 2, to get 8; then add 7, to get 15; then divide that into 30, to get the value for A, which is 2.

Visual Conventions

Figure P-4 shows the conventions that are used in the schematics in this book. A black dot always indicates a connection, except that to minimize ambiguity, the configuration at top-right is avoided, and the configuration at top-center is used instead. Conductors that cross each other without a black dot do not make a connection. The styles at bottom right are sometimes seen elsewhere, but are not used here.

All the schematics are formatted with pale blue backgrounds. This enables components such as switches, transistors, and LEDs to be highlighted in white, drawing attention to them and clarifying the boundary of the component. The white areas have no other meaning.

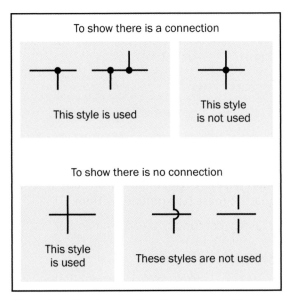

Figure P-4 *Visual conventions that are used in the schematics in this book.*

Units and Backgrounds

So long as the United States clings stubbornly to the habit of expressing dimensions in inches, there is a good argument to follow this custom in books intended for an American audience. With this in mind, Volumes 1 and 2 mostly avoided metric units of length. However, as time passed, the books were translated for use in many parts of the world where the inch is regarded as an anachronism.

Recognizing that we now have an international audience, we have used the metric system throughout this volume (with very few exceptions, such as a photograph of an American plumbing fixture that is designed to fit 3/4" pipe). For readers who are metrically impaired, here are some units of length, and their abbreviations:

- 1 nanometer (nm)
- 1 micrometer (µm) = 1,000nm
- 1 millimeter (mm) = 1,000µm
- 1 centimeter (cm) = 10mm
- 1 meter (m) = 100cm = 1,000mm

A micrometer is also known as a *micron*.

The basic conversion factor from meters to inches is 0.0254. Thus:

- 1 inch = 2.54cm = 25.4mm
- 1/1000 inch = 25.4µm

Sometimes 1/1000 inch is called a *mil*.

In many of the component photographs, a graph-paper background is included. Each square in these backgrounds is 1mm.

To avoid confusion, please remember that a few of these same component photographs appeared in books such as *Make: More Electronics*, where the background grid was in tenths of an inch. Captions to photographs in this volume will remind you that millimeters are now used.

Background colors in the photographs were chosen for contrast with the colors of the components, or for visual variety. They have no other significance.

Component Availability

The world of sensors is changing rapidly, and we have no way of knowing if a component will enjoy a long production run. We recommend checking availability at the following suppliers, which we used frequently during the preparation of the book:

- *http://www.mouser.com*
- *http://www.jameco.com*
- *http://www.sparkfun.com*
- *http://www.adafruit.com*

For obsolete parts, or those that are nearing the end of their commercial life, eBay can be very useful. Alternatively, new substitutions for old parts are often listed at *http://www.mouser.com*.

Issues and Errata

There are three situations where the reader and the writer may want to communicate with each other.

- We may want to tell you if the book contains a mistake of some significance. This is *us-informing-you* feedback.
- You may want to tell us if you think you found an error in the book. This is *you-informing-us* feedback.
- You may be having trouble making something work, and you don't know whether we made a mistake or you made a mistake. You would like some help. This is *you-asking-us* feedback.

Here's how you can deal with each of these situations.

Us Informing You

If you already registered your contact information in connection with *Make: Electronics* (second edition) or *Make: More Electronics*, you don't need to register again for updates relating to the Encyclopedia. If you have not already registered, here's how it works.

The only way you can be notified if there's an error in the book is if you supply your contact information. If we have your email address:

- You will be notified of any significant errors that are found in this book, and you will receive a correction.
- You will be notified if there is a completely new edition of this book, or of *Make: Electronics*, or any other books by Charles Platt. These notifications will be very rare.

Your contact information will not be used for any other purpose.

Simply send a blank email (or include some comments in it, if you like) to:

make.electronics@gmail.com

Please put REGISTER in the subject line.

You Informing Us

If you only want to report an error that you have found, it's really better to use the "errata" system maintained by our publisher. The publisher uses the "errata" information to fix the error in updates of the book.

If you feel sure that you found an error, please visit:

http://bit.ly/encyclopedia_electronic_components_v3

The web page will tell you how to submit errata.

You Asking Us

Our time is obviously limited, but if you have a question, a quick answer may be available. You can send email to *make.electronics@gmail.com* for this purpose. Please put the word HELP in the subject line.

Going Public

There are dozens of forums online where you can discuss this book and mention any problems you are having, but please be aware of the power that you have as a reader, and use it fairly. A single negative review can create a bigger effect than you may realize. It can certainly outweigh half-a-dozen positive reviews.

Responses in the past have been generally positive, but in a couple of cases people have been annoyed over small issues such as being unable to find a part online. Help is available on this kind of topic, if you need it. All you have to do is send a request to *make.electronics@gmail.com*.

Safari® Books Online

 Safari Books Online is an on-demand digital library that delivers expert content in both book and video form from the world's leading authors in technology and business.

Technology professionals, software developers, web designers, and business and creative professionals use Safari Books Online as their primary resource for research, problem solving, learning, and certification training.

Safari Books Online offers a range of plans and pricing for enterprise, government, education, and individuals.

Members have access to thousands of books, training videos, and prepublication manuscripts in one fully searchable database from publishers like O'Reilly Media, Prentice Hall Professional, Addison-Wesley Professional, Microsoft Press, Sams, Que, Peachpit Press, Focal Press, Cisco Press, John Wiley & Sons, Syngress, Morgan Kaufmann, IBM Redbooks, Packt, Adobe Press, FT Press, Apress, Manning, New Riders, McGraw-Hill, Jones & Bartlett, Course Technology, and hundreds more. For more information about Safari Books Online, please visit us online.

You can access the errata page at *http://bit.ly/ encyclopedia-electronic-components-v3*.

Make: unites, inspires, informs, and entertains a growing community of resourceful people who undertake amazing projects in their backyards, basements, and garages. Make: celebrates your right to tweak, hack, and bend any technology to your will. The Make: audience continues to be a growing culture and community that believes in bettering ourselves, our environment, our educational system—our entire world. This is much more than an audience, it's a worldwide movement that Make is leading. We call it the Maker Movement.

For more information about Make:, visit us online:

Make: magazine: *http://makezine.com/magazine*
Maker Faire: *http://makerfaire.com*
Makezine.com: *http://makezine.com*
Maker Shed: *http://makershed.com*

To comment or ask technical questions about this book, send email to:

bookquestions@oreilly.com.

Acknowledgments

Datasheets and tutorials maintained by component manufacturers were considered the most trustworthy sources of information online. In addition, component retailers, college texts, crowd-sourced reference works, and hobbyist sites were used. The following books provided useful information:

Boylestad, Robert L. and Nashelsky, Louis: *Electronic Devices and Circuit Theory*, 9th edition. Pearson Education, 2006.

Braga, Newton C.: *CMOS Sourcebook*. Sams Technical Publishing, 2001.

Hoenig, Stuart A.: *How to Build and Use Electronic Devices Without Frustration, Panic, Mountains of Money, or an Engineering Degree*, 2nd edition. Little, Brown, 1980.

Horn, Delton T.: *Electronic Components*. Tab Books, 1992.

Horn, Delton T.: *Electronics Theory*, 4th edition. Tab Books, 1994.

Horowitz, Paul and Hill, Winfield: *The Art of Electronics*, 2nd edition. Cambridge University Press, 1989.

Ibrahim, Dogan: *Using LEDs, LCDs, and GLCDs in Microcontroller Projects*. John Wiley & Sons, 2012.

Kumar, A. Anand: *Fundamentals of Digital Circuits*, 2nd edition. PHI Learning, 2009.

Lancaster, Don: *TTL Cookbook*. Howard W. Sams & Co, 1974.

Lenk, Ron and Lenk, Carol: *Practical Lighting Design with LEDs*. John Wiley & Sons, 2011.

Lowe, Doug: *Electronics All-in-One for Dummies*. John Wiley & Sons, 2012.

Mims III, Forrest M.: *Getting Started in Electronics*. Master Publishing, 2000.

Mims III, Forrest M.: *Electronic Sensor Circuits & Projects*. Master Publishing, 2007.

Mims III, Forrest M.: *Timer, Op Amp, & Optelectronic Circuits and Projects*. Master Publishing, 2007.

Predko, Mike: *123 Robotics Experiments for the Evil Genius*. McGraw-Hill, 2004.

Scherz, Paul: *Practical Electronics for Inventors*, 2nd edition. McGraw-Hill, 2007.

Williams, Tim: *The Circuit Designer's Companion*, 2nd edition. Newnes, 2005.

In addition, three individuals provided special assistance. Our editor, Brian Jepson, was immensely helpful in the development of this book. Philipp Marek reviewed the text for errors, and Erico Narita collaborated on the Photoshop work.

GPS

The acronym **GPS** properly refers to the entire Global Positioning System, including satellites and ground-based control installations. However, a *GPS sensor* consists of a surface-mount chip that processes signals from GPS satellites using a small rectangular antenna, often mounted on top of a *GPS chip*.

A *GPS module* is usually a small board on which a GPS sensor is mounted with additional components. A *GPS receiver* is a device including a data display and other features, such as memory, in addition to a GPS module. In casual colloquial usage, someone who refers to "a GPS" usually means a GPS receiver.

GPS is almost always capitalized without periods.

OTHER RELATED COMPONENTS

- **magnetometer** (see Chapter 2)

What It Does

The Global Positioning System is a navigational aid jointly funded by the U.S. Department of Defense and the U.S. Department of Transportation, while being maintained by the U.S. Air Force. Signals from GPS satellites can be received and processed by modules in a wide variety of equipment ranging from aircraft to wristwatches. The signals provide location data, and may also be used as an accurate time reference.

Schematic Symbol

There is no specific schematic symbol for a GPS chip. It is likely to be shown as a box containing abbreviations that define pin functions, similar to any integrated circuit chip.

GPS Segments

The Global Positioning System consists of three segments:

The space segment

This originally required 24 communications satellites, but was revised in 2011 to require 27, to provide better global coverage. As of August 2015, there were actually 31 satellites in service, with additional "spares" that can be activated if necessary. The satellites occupy orbits 12,500 miles above the Earth, allowing each of them to circle the planet twice in 24 hours. Specifications are maintained online (*http://www.gps.gov/technical/ps/*).

The control segment

This includes a master ground-based control station, an alternate master control station, 12 command and control antennas, and 16 monitoring sites, all maintained by the U.S. Air Force.

The user segment
> This consists of receiving devices, including those that are government-owned and those that are privately owned.

How It Works

Each satellite carries multiple atomic clocks that maintain precise time, and a pseudo-random number generator in the form of a linear-feedback **shift register** (see Volume 2).

A GPS receiver can distinguish the signals from at least four satellites by comparing their received pseudo-random bit sequences, and can compute the receiver's distance to each satellite by comparing the arrival times of satellite signals.

When a satellite appears above the horizon, it approaches a receiver. After passing overhead, it moves away. This relative motion causes a *Doppler shift* in the received frequency, which the receiver circuit must take into account.

GPS satellites transmit on several frequencies simultaneously. The one for civilian use is 1575.42 MHz, called L1. Another one, 1227.6 MHz, called L2, is reserved for military use.

Variants

A GPS chip generally processes input from an antenna and provides output through solder pads. The antenna is often integrated as a ceramic square or rectangle mounted above the chip, but many chips can also process input from an external antenna. Figure 1-1 shows a GPS chip with a metal shield that is easily mistaken for an antenna. In Figure 1-2, the GPS sensor does incorporate a ceramic antenna.

Some GPS chips contain flash memory for internal data logging, although this is not a standard feature.

Suppliers such as Adafruit and Sparkfun offer GPS modules mounted on breakout boards for easier connection with other components, as shown in Figure 1-2. Some breakout boards also include provision for battery backup with a button cell.

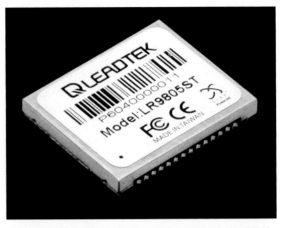

Figure 1-1 *A GPS sensor. This surface-mount chip is hidden beneath a metal shield.*

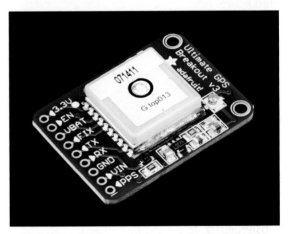

Figure 1-2 *A breakout board offered by Adafruit, incorporating a GPS sensor.*

GPS capability is almost always included in modern smartphones and tablets. It is used in handheld devices for navigation on foot, and in devices designed to be mounted in motor vehicles. Cars offer GPS capability as an option accessed via a built-in screen.

A *GPS tracker* is a device that may lack a display and simply logs its position to internal memory, from which data can be downloaded to a computer later. Many (older) handheld GPS receivers have a connector giving access to a serial or USB port, and provide data in the same NMEA format as the GPS modules described below.

After the Global Positioning System became widely used, other competing systems were introduced. These include the European Galileo, the Russian GLONASS (an acronym for Global Navigation Satellite System), and the Chinese Beidou. As of 2015, GLONASS had become fully operational. Some receivers, including those in cellular phones, compare signals from GPS and GLONASS satellites to achieve higher accuracy.

Values

Sensitivity is expressed usually in dBm, meaning the power ratio in decibels (dB) of the measured power referenced to one milliwatt (mW).

Time to First Fix (TTFF) is the time required to obtain an initial fix from a satellite.

Number of channels is the number of satellites that a GPS receiver can track simultaneously. Early GPS receivers could sense only four channels. Modern units may be able to deal with 22.

Power consumption may be measured in milliwatts. For example, the FGPMMOPA6H GPS standalone module by G.top claims power consumption of 82mW during acquisition of satellite signals and 66mW while tracking subsequently. At a typical voltage of 4VDC, the chipset consumes about 20mA and approximately 17mA, respectively.

Form factor. This is the size of a chip, often determined by the dimensions of the ceramic antenna on top of it. Dimensions may be 15mm x 15mm or larger.

Update rate. The number of position measurements per second. While 1 update per second is often sufficient, some chips generate updates more rapidly. The frequency of updates is expressed in hertz.

Output type. This is often TTL-level serial providing NMEA data. The baud rate can vary and is often selectable.

Supply voltage. Often below 5VDC.

Current consumption. Higher during satellite acquisition.

How to Use It

A GPS module requires only a DC power supply, and will start outputting data as soon as it has identified satellites that are currently within range.

Data provided by a GPS module uses a rather slow and primitive plain-ASCII protocol known as NMEA, developed by the National Marine Electronics Association. Each block of data is known as a *sentence*, and can be parsed independently of previous and subsequent sentences. The default transmission rate is 4,800 bits per second, using 8 bits to identify an ASCII character, no parity, and 1 stop bit. However, some GPS modules use a serial rate of 9,600 bps or faster.

A sentence begins with a two-letter abbreviation defining the type of device employing the sentence. For a GPS device, the abbreviation is GP. The sentence continues with another abbreviation of three characters or more, describing the type of data being transmitted, so that numeric values in the sentence can be interpreted correctly.

The remainder of a sentence consists of letters and numerals in plain ASCII, with values separated by commas. A sentence cannot contain more than 80 characters. A sentence will specify the latitude, longitude, and altitude of the GPS, and a value defining the time when the readings were derived from satellite signals. Some sentence data structures are proprietary, established by the device manufacturer, and will begin with the letter P.

A GPS device may send a variety of different sentence types in succession, to overcome the 80-character sentence length limit. Each sentence will be preceded with an identifier. Sentence types and data contents will be defined in a manufacturer's datasheet.

Output from GPS chips may be compatible with a microcontroller. Output from a GPS breakout board will almost certainly be compatible, and the board is likely to have its own voltage regulator. The microcontroller can receive serial data from the GPS chip, can stop the GPS chip via its enable pin, and can start and stop internal logging of data in flash memory on the GPS chip if this feature is available.

Code libraries are available online for a microcontroller, to enable it to receive and interpret serial data from a GPS device.

Pulse per Second Output

As the GPS positioning depends on calculating distances from the time it takes for a radio signal to travel, accurate timekeeping is needed. When a GPS receiver obtains a fix for its position, it will get a value for the current time as well. This makes GPS receivers usable to supply time and frequency standards. Most receiver modules report the time together with the position. Many also provide a special PPS output, which produces one pulse per second, precisely synchronized with the satellites in view.

The accurate time provided by the GPS receiver can be used to *discipline* a crystal oscillator. This means measuring the crystal oscillator frequency relative to a GPS-provided reference, and continuously adjusting the crystal frequency to keep it stable.

What Can Go Wrong

Problems generally affect chips and modules rather than devices.

Electrostatic Discharge

The patch antenna on a GPS chip connects with the chip via an RF input. If the antenna is subjected to an electrostatic discharge, the chip can be permanently damaged. Likewise, damage will result if a discharge is applied to the RF input, for example through a soldering iron. The chip must be grounded before any work is done involving the RF input.

Failure to Ground Properly

The ground solder pad on a chip, or the ground pin on a breakout board, must make contact before voltage is applied to other pads or pins.

Cold Joints

The patch antenna on a GPS chip can function as a heat sink, increasing the risk of cold solder joints when the chip is being mounted on a board.

Restricted Availability

U.S. regulations limit the export of some GPS devices capable of rapid positional updates that could be usable in military aircraft or missiles. Other restrictions also apply. Suppliers may be inconsistent in their refusal to allow purchase of restricted items outside of the United States.

Inability to Detect Satellites

Any GPS device may fail to detect satellite signals if its view of the sky is obstructed. Reception is usually possible through window glass, but may not be available through walls, a roof, heavy tree cover, and other natural obstructions.

Exceeding Maximum Velocity or Altitude

Security regulations limit the capability of GPS devices to function above 60,000 feet or at greater than 1,200 miles per hour. Outside of these limits, the GPS device will not provide data. This may affect applications in amateur rocketry or high-altitude balloons.

magnetometer

2

This entry deals only with magnetic sensors that respond to the Earth's magnetic field. Small magnetic sensors such as the ubiquitous *Hall sensor* may be used for many other purposes, such as determining the position or rotational speed of mechanical components. These applications involve **object presence** sensing; see Chapter 3. In that entry, Hall sensors are discussed at "Hall-Effect Sensor".

In the past, a **magnetometer** was a bulky measurement device incorporating knobs or other controls and some form of display. While that use of the word is still common, this entry deals only with chip-based magnetic sensors.

OTHER RELATED COMPONENTS

- **accelerometer** (see Chapter 10)
- **gyroscope** (see Chapter 9)
- **GPS** (see Chapter 1)

What It Does

A traditional compass consists of a thin magnetized strip of metal balanced on a pivot. It aligns itself with the Earth's magnetic field.

A *scalar magnetometer* measures the total strength of a magnetic field. A *vector magnetometer* can measure the strength in a specified direction. In particular, it can provide a numeric output describing the angle between the orientation of the measuring device and the Earth's magnetic poles.

Chip-based magnetometers are usually vector-type, containing three sensors mounted orthogonally—that is, each of them at 90 degrees to the other two. Suitable software can interpret the analog readings from the sensors to calculate magnetic north or south regardless of the angle at which the instrument is being held, relative to the ground.

Schematic Symbol

There is no specific schematic symbol for a magnetometer.

IMU

A **gyroscope** measures the rate of rotation of the enclosure in which it is mounted. This is properly known as the *angular velocity*. A gyroscope will also respond to changes in the rate of rotation. It does not measure linear motion or a static angle of orientation.

An **accelerometer** measures variations in linear motion and will also measure its own static orientation relative to the force of gravity. If an accelerometer rotates around its own axis, it will not measure angular velocity.

When an accelerometer and a gyroscope are contained in the same package, optionally with a magnetometer, they may be described as an *IMU* (inertial measurement unit), which can provide necessary data to maneuver aircraft, spacecraft, and watercraft, especially when **GPS** signals are unavailable.

Applications

Magnetometers are found in handheld equipment such as digital compasses, cameras, and cellular phones. They are usually surface-mount chips that are manufactured in large quantities and may be used in conjunction with microcontrollers. For the hobby-electronics community, or for experimental product development, a magnetometer may be mounted on a breakout board for ease of use. A board using a Honeywell HMC5883L is shown in Figure 2-1.

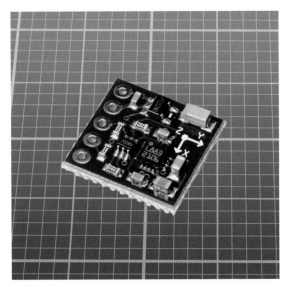

Figure 2-1 *The Honeywell HMC5883L 3-axis magnetometer mounted on a breakout board. The background grid is in millimeters.*

How It Works

An explanation of magnetometers requires an understanding of the fundamentals of magnetism.

Magnetic Fields

A magnetic field is often represented by *field lines* that show the strength and vector of the field. Field lines associated with a simple permanent magnet are shown in Figure 2-2, where the strength, or *flux density*, at any point, is indicated by the spacing the lines, while the angle tangential to a line indicates its vector. (For an extended discussion of magnetism, see **electromagnet** in Volume 1.)

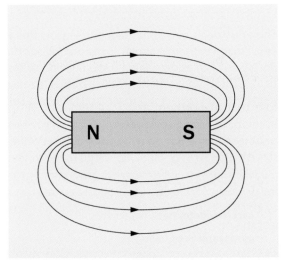

Figure 2-2 *Field lines representing magnetic flux created by a bar magnet. Space between the field lines is inversely proportional to flux density. In reality this is a three-dimensional effect, and a more accurate representation would show the magnet and field lines revolved around the axis of the magnet.*

The flux density of a magnetic field is usually represented by the letter B, and is measured in *newton-meters per ampere*, more commonly referred to as *teslas* (T). An older unit of measurement was gauss (G), 1 tesla being equivalent to 10,000 gauss. Some datasheets still refer to gauss.

The Earth's magnetic field is believed to result from convection currents in the outer liquid of the Earth's core. The strength of this field varies from 25 to 65 microteslas (0.25 to 0.65 gauss) depending on the location where it is measured. To a very rough approximation, the Earth behaves as if a giant bar magnet connects the magnetic north pole and the magnetic south pole. See Figure 2-3.

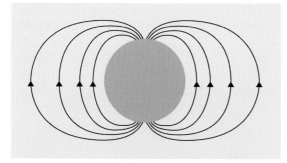

Figure 2-3 *The magnetic field of the Earth approximately resembles the field around a bar magnet.*

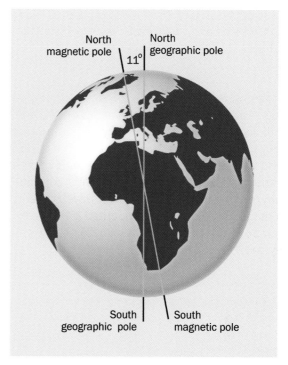

Figure 2-4 *The angle between the magnetic axis and the rotational axis of the Earth is approximately 11 degrees.*

Note that in northern and southern latitudes, the field lines angle steeply down toward the surface, while the lines are approximately parallel with the surface near the equator. Consequently a magnetometer held tangentially to the surface of the Earth will tend to measure a stronger horizontal field near the equator than near the poles.

The varying angle of the field lines tangential to the Earth is known as their *inclination*. Variations in field strength can be used to determine an approximate location, although the **GPS** (Global Positioning System, using satellites) enables this much more accurately.

Confusion results from the fact that the nothern magnetic pole of the Earth actually behaves as a south pole, while the southern magnetic pole behaves as a north pole. That is, when a permanent magnet is free to rotate, its north pole will orient itself toward the so-called northern magnetic pole of the Earth, even though opposite poles attract. The northern pole of the Earth should therefore be thought of as the pole that attracts the north end of a compass.

Earth's Axes

Planet Earth revolves around an imaginary line referred to as the *axis of rotation*. This line is close to, but not quite coincident with, the *magnetic axis* connecting the magnetic north pole and magnetic south pole, as shown in Figure 2-4.

Magnetic declination is the angle between the magnetic north pole and the geographical north pole, as perceived by an observer. This angle will vary depending on the observer's location on the surface.

Because of declination, the direction of magnetic force at points on the surface of the planet will vary with latitude and longitude, as shown in Figure 2-5. In this figure, the *magnetic meridians* are shown in red, superimposed on the *geographical meridians* shown in green. Magnetic meridians show the direction of magnetic force, while geographic meridians are drawn between the ends of the axis of rotation of the Earth. While there is an approximate correlation between the two, in some areas, especially near the north and south poles, the discrepancy can be more than 40 degrees.

Figure 2-5 *The red lines indicate the direction in which a compass would be likely to point to magnetic north. The green lines connect the geographical poles of the planet. (From Wikimedia Commons.)*

Standard declination tables for locations on the Earth are available, and these values must be added to, or subtracted from, the reading from a compass or magnetometer to determine the direction of geographical north. Navigational systems customarily express the heading of a vehicle or vessel relative to geographical north, as shown in Figure 2-6.

Coil Magnetometer

Current flowing through a wire creates a magnetic field with flux density that is directly proportional to the current in amperes. Conversely, a changing magnetic field will induce current in a wire. This principle is used in a *coil magnetometer*, which can detect buried objects when the coil moves above them. A *rotating coil magnetometer* can determine magnetic field strength while remaining in a stationary position. However, the coil in a coil magnetometer must be relatively large.

Hall Effect and Magnetoresistance

In modern handheld devices, a magnetometer will generally use either the *Hall effect* (see

"Hall-Effect Sensor") or *magnetoresistance*, described here.

Figure 2-6 *Heading is an angle usually calculated relative to geographical north, sometimes referred to as "true north."*

Magnetoresistance is a phenomenon where the resistance of a material changes fractionally when it is exposed to a magnetic field. This is capable of yielding greater accuracy than Hall-effect sensors, but has tended to be more expensive.

Orthogonally oriented sensors contained in a surface-mount chip are aligned with axes identified with letters X, Y, and Z. These sensors are analog devices whose values are converted to digital values by an onboard *analog-to-digital converter* (ADC). The values are stored in *registers* that are available to other devices, often via the *I2C* protocol, which is widely used by microcontrollers.

Typically there will be two eight-bit registers for each axis, one defining the high byte and the other defining the low byte for the digital value. In reality the ADC is likely to use 10 to 13 bits, with the remaining 6 to 3 bits being unused.

Variants

The Freescale Semiconductor FXMS3110 is a typical low-cost chip containing a 3-axis magnetometer sensor. Many chips now additionally include a 3-axis accelerometer sensor. An example is the LSM303 by STMicroelectronics, which is sold on a breakout board by Adafruit. See Figure 2-7.

Figure 2-7 *The LSM303 is a chip manufactured by STMicroelectronics. It is shown here on a breakout board from Adafruit.*

See Chapter 10 for an explanation of the function of accelerometers.

The InvenSense MPU-9250 is a highly sophisticated IMU, including a 3-axis **gyroscope** in addition to a 3-axis magnetometer and 3-axis accelerometer.

A processing unit in the MPU-9250 reconciles the nine variables, and the digital output can be accessed via I2C or SPI protocol at speeds up to 1MHz. All functions of the chip are contained within a package measuring less than 3mm square.

See Chapter 9 for additional details about the functioning of a gyroscope.

How to Use It

A basic 3-axis magnetometer sensor such as the HMC5883L can be tested with a microcontroller that can receive data from its registers via I2C protocol. This is relatively easy with an Arduino, which was designed to be I2C-capable.

Several breakout boards are available with the HMC5883L mounted on them. Many of these boards contain a voltage regulator, allowing a 5VDC power supply to be used even though the chip is designed for a typical supply of 2.5VDC.

In addition to the power supply, the breakout board only requires two connections for I2C communication, to its SCL pin (serial clock input) and SDA pin (serial data input/output). If basic I2C software is installed on the microcontroller, it will read digital values from the magnetometer registers. If additional software is used, it will convert the values to magnetic flux densities in microteslas for the X, Y, and Z axes. Code libraries to achieve these goals are widely available online.

A more sophisticated chip such as the InvenSense MPU-9250 can be used similarly, but yields additional data for conversion. Here again, code libraries can be found online. The slightly older MPU-9150 is sold on a breakout board with downloadable code from Sparkfun.

What Can Go Wrong

Bias

Magnetometers are sensitive to their environment, which can induce *magnetometer bias* of two types.

Hard-iron bias is primarily caused by magnetized material inside the device containing the magnetometer. Because this effect is usually unvarying, compensating for it is relatively easy. *Soft-iron bias* is caused by interaction between variations in the Earth's magnetic field and materials inside a magnetometer that can be magnetized.

A common example of soft-iron bias would be power lines, generating a magnetic field that can affect model aircraft and drones using magnetometers for navigation.

Mounting Errors

Placement of a chip-based magnetometer on a circuit board is critical. The field effects of transformers or relays must be taken into account, and even the low voltage and low current in a circuit trace can create a magnetic field sufficient to disturb a chip. No traces on any layer of the board should pass across the footprint of the chip. A magnetometer should not be mounted within a ferromagnetic case.

object presence sensor

An **object presence** sensor may also be described as an *object detector* or *detection sensor*.

The term **proximity** sensor may be applied to this component. However, in this Encyclopedia a proximity sensor has the capability of estimating the distance to a target. See Chapter 5. An object presence sensor merely detects whether an object is within a preset range, and does not provide additional information.

Optical and *magnetic* object-presence sensors are described and compared in this entry. *Ultrasonic sensors* are described in the entry dealing with proximity sensors, because they tend to be used for distance measurement rather than just for detection. Other methods of object presence sensing, including *capacitive, doppler, inductive, radar,* and *sonar,* are not included in this Encyclopedia.

A sensor that detects an object by receiving light reflected from it is categorized as a *reflective* sensor, and is included in this entry. (If a module includes a light source as well as a light sensor, it is properly categorized as *retroreflective,* although that term is not always used.)

A sensor that detects an object when it interrupts a beam of light is a *transmissive* sensor, and is included in this entry. It may also be described as a *through-beam* sensor or as an *optical switch.*

A sensor that responds to the motion of an object that emits infrared radiation is a **passive infrared** motion sensor, also described by its acronym, PIR, and often referred to simply as a *motion sensor.* It has its own separate entry. See Chapter 4.

Phototransistors and **photodiodes** may be used as sensing elements in presence sensors. These components have their own entries as light sensors. See Chapter 22 and Chapter 21.

OTHER RELATED COMPONENTS

- **proximity** sensor (see Chapter 5)
- **passive infrared** sensor (see Chapter 4)

What It Does

An **object presence** sensor verifies the presence or absence of an object within a predetermined range, without necessarily measuring how far away it is or how fast it is moving. The object may be described as a *target.*

Object presence sensing is often used to verify the correct function of an automated system—for example, the placement of objects on a conveyor belt. It can also be used to count objects as they pass a sensor.

Some types of security systems use presence sensors to sound an alarm if an intruder interrupts a beam of light. They can verify that a door or window is closed. They may also function as a limit switch to control the operation of a motor.

Schematic Symbol

In schematics, an optical presence sensor may be indicated with the symbol for an LED, plus the symbol for a **phototransistor**, with one or two arrows connecting them, as shown at top-left in Figure 3-1. Wavy-line arrows may indicate an infrared connection.

A **photodiode** may be substituted for a phototransistor, as shown at top-right in the figure.

A magnetic sensor may be shown with the symbol for a *Hall-effect sensor*, as in the lower half of Figure 3-1.

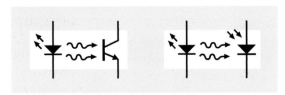

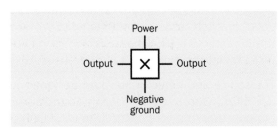

Figure 3-1 *Top: two possible schematic symbols for an optical presence sensor, using a phototransistor (left) or photodiode (right). Other variants are possible. Bottom: the schematic symbol for a Hall-effect sensor, commonly used in a magnetic sensor.*

Variants

To assist the reader in comparing different options for detecting the presence of an object, this entry includes two primary variants: *optical* and *magnetic*.

The optical sensors are divided into *transmissive* and *reflective* (including *retroreflective*). The magnetic sensors are *reed switches* and *Hall-effect sensors*. A chart showing the categories and subcategories is shown in Figure 3-2.

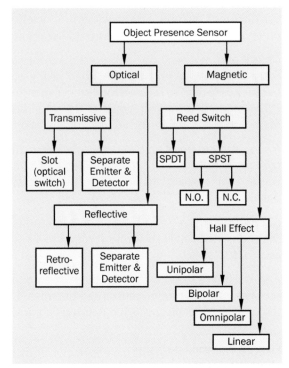

Figure 3-2 *Categories of object presence sensors discussed in this entry. Other types of object presence sensors exist, but are less common, and are not included here.*

Optical Detection

A *transmissive optical sensor*, also known as a *through-beam sensor*, is really a pair of components, one emitting light and the other receiving it. The sensor is triggered if an object interrupts or reflects the light beam, as shown in Figure 3-3.

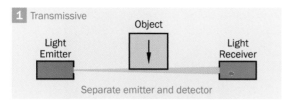

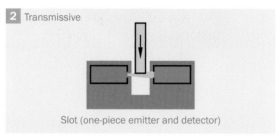

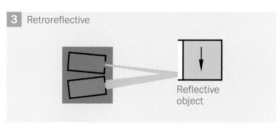

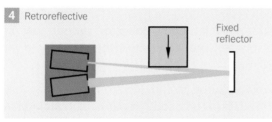

Figure 3-3 *Various configurations of optically activated object presence sensors. See text for details.*

If the light emitter and the light detector face each other across a small gap, they may be contained in a single module (usually with a slot in it) as shown in section 2 of Figure 3-3. This is often referred to as an *optical switch* (not to be confused with the solid-state switching devices used in telecommunications). It is sometimes described as a *photointerruptor* or *photointerrupter*.

A *reflective optical sensor* also consists of a light emitter and a light detector, but they are placed adjacent to each other, facing in the same general direction. When they are mounted in one module, as is often the case, the arrangement is properly known as a *retroreflective sensor*. Either way, the combination may be triggered in one of two ways:

- An object passing in front of the light beam reflects it back to the detector. The object must be naturally reflective (for example, glass containers or white boxes on a conveyor), or must have a reflective patch applied to it, or the light source must be bright enough to reflect from an object that is not highly reflective. This configuration is shown in section 3 of Figure 3-3.

- A stationary reflector can be mounted opposite the light emitter, in which case a detector beside the light emitter is triggered when an object interrupts the reflected light beam. This configuration is shown in section 4 of Figure 3-3.

Transmissive Optical Sensors

A light source and light detector may be sold as separate components in a matched pair. An example is the Vishay TCZT8020, shown in Figure 3-4. These components are small, each measuring no more than 5mm x 3mm. They are designed to be placed just a few millimeters apart. The light source is an infrared LED, while the detector is a **phototransistor** (see Chapter 22 for information about phototransistors).

The source and detector are both designed to use 5VDC. Output from the phototransistor is an open collector. Current through the collector must not exceed 50mA, and must be limited by a pullup resistor of 100 ohms or more. Current through the source must not exceed 60mA, and must be limited by an appropriate series resistor.

Details on using an open-collector output will be found in the Appendix. See Figure A-4.

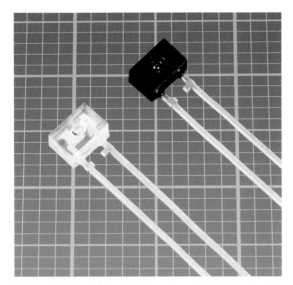

Figure 3-4 *A matched light source and light detector, for use as a transmissive optical sensor. The background grid is in millimeters.*

Figure 3-5 *A low-cost transmissive sensor, more commonly known as an optical switch. The background grid is in millimeters. The slot in the sensor is about 5mm wide.*

The Omron EE-SX range includes a variety of configurations of light source and detector separated by a 5mm slot in one module. The source is an infrared LED, and the detector is a phototransistor.

The Omron components tolerate a wide range of supply voltages, from 5VDC to 24VDC, with no series resistor necessary for the LED. Open-collector output from the phototransistor can tolerate 50mA to 100mA, depending on the particular version of the sensor. A red indicator LED is illuminated when an object blocks the slot in the sensor. Some versions have a high output when the slot is open, while others have a high output when the slot is closed. Because of their various features, these sensors are relatively expensive.

A much cheaper optical switch is the Everlight ITR9606 (described by the manufacturer as an "opto interrupter"). It is pictured in Figure 3-5. This is intended as a 5V device and has an open-collector output. It requires a series resistor for the LED and a pullup resistor for the open-collector output. Many similar detectors are available.

For longer-distance detection, an infrared receiver can be mounted separately from an infrared LED. The TSSP77038 from Vishay detects infrared light from as far away as 50cm, and pulls its open collector output low in response. The light must be modulated at 38kHz.

Polulo Robotics and Electronics sells a very affordable breakout board containing a TSSP77038 receiver paired with an infrared LED modulated by a 555 timer. It is shown in Figure 3-6. Because this board contains a light source as well as a light detector, it is really a retroreflective sensor.

Where distances exceeding 1 meter are involved, a laser coupled with a phototransistor that is shielded from ambient light may be necessary.

Figure 3-6 *The Vishay TSSP77038 installed with an appropriate light source on a breakout board from Pololu Robotics and Electronics.*

Retroreflective Optical Sensors

As is the case with a transmissive object detector, the retroreflective type may be listed by vendors as an *optical switch*. Other terms used in datasheets include *reflective interrupter*, *reflective object sensor*, *reflective optical sensor*, *reflective photointerrupter*, *opt-pass sensor*, and *photomicrosensor (reflective)*. The remarkable lack of standardization in terminology creates a problem when searching online for these devices. Why so many different names evolved for the exact same component is unclear.

Many retroreflective object-detection sensors are available in packages ranging from 5mm x 5mm to 10mm x 10mm in size. Almost all of these modules are analog devices using an infrared LED as the light source and a phototransistor as the sensor, with an open-collector output. (For more information about phototransistors, see Chapter 22.)

With a suitable pullup resistor, the output voltage will be proportional with the inverse of the distance. If V is the voltage, d is the distance, and k is a conversion factor:

```
V = k * ( 1 / d)
```

While many of the smaller modules are surface-mount, some have leads, as shown in Figure 3-7. A major limitation of these small components is that they have a very limited

range, typically less than 5mm. They can only be used in applications such as process control where the position of a target is controlled and predictable.

Figure 3-7 *The Rodan RT-530 is a small object-presence sensor with a limited range typical of this type of retroreflective component. The background grid is in millimeters.*

Another example of a retroreflective sensor in a small package is the Optek OPB606A shown in Figure 3-8. The background grid is in millimeters.

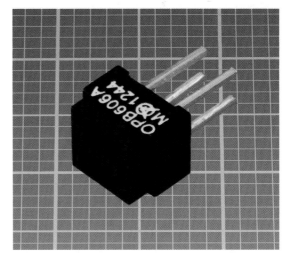

Figure 3-8 *The Optek OPB606A. The background scale is in millimeters.*

A retroreflective module, with a lensed LED and lensed phototransistor to focus the outgoing and reflected beams, is the Vishay TCRT5000 shown in Figure 3-9.

Figure 3-9 *The Vishay TCRT5000 retroreflective sensor. The background grid is in millimeters.*

Retroreflective modules that have a greater range tend to be physically larger, less common, and more expensive. Sharp makes a popular series. Some part numbers and distance sensing limits are: GP2Y0D805Z0F (5mm to 5cm), GP2Y0D810Z0F (2cm to 10cm), and GP2Y0D815Z0F (5mm to 15cm). Figure 3-10 shows the GP2Y0D810Z0F mounted on a small breakout board from Polulo Robotics and Electronics. The board is useful, as the pins on the sensor are spaced only 1.5mm apart. The size of the board is approximately 8mm x 20mm.

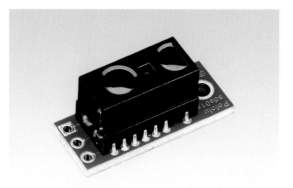

Figure 3-10 *The Sharp GP2Y0D810Z0F detection sensor mounted on a board from Polulo Robotics and Electronics. From a photograph by Adafruit Industries.*

Each of the detection sensors in this series is described in Sharp datasheets as a "Distance Measuring Sensor Unit," but in fact they do not measure distance. A single output is normally logic-high and drops to logic-low when a target is within range. Sharp refers to this as a "digital" output, but in fact it is binary, and should not be confused with the digital buffer on a more sophisticated proximity sensor that contains an analog-to-digital converter and provides a numeric output.

It is important to distinguish the Sharp object-presence sensors listed above from the range of Sharp **proximity** sensors described in the entry in Chapter 5. The proximity sensors are physically larger, and most have an analog output that varies with the distance of a target.

Magnetic Sensors

Prepackaged magnetic sensing units are sold in many configurations for industrial and military applications. Although they may be referred to as "magnetic sensors," they are outside the scope of this Encyclopedia. Here we discuss board-mounted components. Almost always, they use a *reed switch* or a *Hall-effect* sensor as their sensing elements.

Reed Switch

A reed switch is a magnetically activated mechanical switch. It consists of two metallic contacts in a small enclosure that is usually a glass capsule. The contacts are magnetic, and move in response to a magnetic field. A permanent magnet is used to activate the switch. Two reed switches are shown in Figure 3-11.

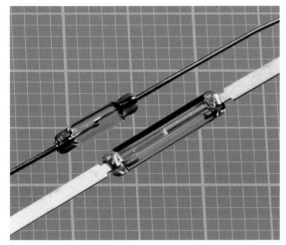

Figure 3-11 *Two SPST reed switches. Although the contacts may seem to be touching, in fact they are separated by a tiny gap, and these switches are the normally open type. The background grid is in millimeters.*

A reed switch exhibits a small amount of *hysteresis*, because the magnetic field strength required to overcome the mechanical resistance of the springy contacts is greater than the field strength required to keep them closed.

Very small electromagnetic relays that only switch a very low current may use a coil-activated reed switch. For purposes of this Encyclopedia, such a component is considered to be a relay, not a sensor. For more information about **relays**, see the entry in Volume 1.

The most common everyday application for a reed switch is in an alarm system that is triggered by an intruder entering a building. A magnet in a sealed plastic enclosure is attached to a door or window, while a reed switch, in another sealed plastic enclosure, is attached to the frame, very close to the magnet. Typical components of this type are shown in Figure 3-12. A diagram illustrating the mode of operation appears in Figure 3-13.

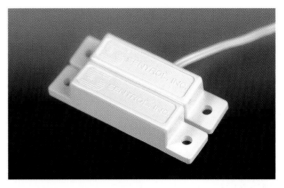

Figure 3-12 *A typical alarm sensor to detect the opening of a door or window. The nearer module contains a magnet, while the module behind it contains a reed switch.*

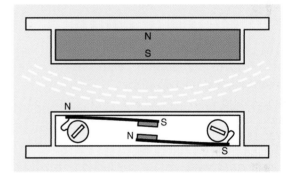

Figure 3-13 *The dashed white lines suggest the magnetic field that closes the contacts of a reed switch.*

So long as the alarmed door or window remains closed, the magnet activates the reed switch. If the door or window is opened, the magnet moves away from the reed switch, and its contacts relax. Usually in this application the reed switch is the normally open type, and is held in its closed state by the magnet. This allows multiple switches to be wired in series, completing a circuit. If a single switch opens, the circuit is broken, and an alarm is triggered.

Reed Switch Variants

Most reed switches are SPST, either normally open or normally closed, although normally open variants are more common. Some switches are SPDT, although this variant is relatively rare. An example of a SPDT reed switch is shown in Figure 3-14.

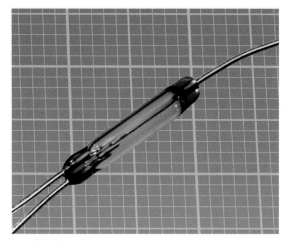

Figure 3-14 *A SPDT reed switch. The background grid is in millimeters.*

The physical size of a reed switch tends to vary roughly in proportion with its current-switching capability.

Reed switches are most commonly supplied with axial leads. A minority are available for surface-mount applications.

Some reed switches are packaged in a plastic pod to provide physical protection for the glass capsule.

Reed Switch Values

The datasheet for a reed switch is likely to contain the following values:

Pull-in: The minimum magnetic field strength required to activate the switch, often measured in ampere-turns.

Drop-out: The maximum magnetic field strength that allows the contacts of the switch to relax, often measured in ampere-turns. Pull-in will be a higher value than drop-out.

Maximum switching current: While a few industrial reed switches can switch as much as 100A, they are uncommon and expensive. A common

value for a reed switch about 15mm long is 500mA.

Maximum carry current: If specified, it will be higher than the switching current.

Maximum switched power: Because a reed switch can be used with alternating current, its switching capability may be expressed in watts, or as VA (volts multiplied by amps). 10VA is a common value.

Maximum voltage: While reed switches are most often used with low voltages, some are rated to switch up to 200V.

How to Use a Reed Switch

While an optical object presence sensor may be supplied in a single package containing a light emitter and a light detector, a reed switch always requires an activating magnet that is mounted separately. For reliable operation, the maximum distance between the switch and the magnet is usually restricted to a few millimeters.

The orientation of a magnet that activates a reed switch is not crucial, but will affect the sensitivity of the switch. A manufacturer's datasheet should be consulted for information about optimal magnetic polarity.

As in other mechanical switches, the contacts in reed switches vibrate very briefly when opening or closing. This is known as *contact bounce*, and can be misintepreted as a series of separate signals by digital logic or a microcontroller. *Debouncing* with hardware designed for that purpose, or software (in a microcontroller), may be necessary. See the entry on **switches** in Volume 1.

Hall-Effect Sensor

A Hall-effect sensor responds to a magnetic field by generating a small voltage, usually amplified by a transistor included in the package with the sensor.

When a Hall sensor is in its "off" state (i.e., not triggered by a magnetic field), the sensor creates high resistance between the collector of an internal NPN transistor and negative ground. Consequently, if voltage through a pullup resistor is applied to the collector, the voltage on the collector will be high.

When the sensor is in its "on" state, the resistance drops, the voltage supplied to the collector from the pullup resistor is shunted to ground, and the output voltage appears to go low. As a general rule:

- Activated Hall sensors appear to have a low output.

- Inactive Hall sensors appear to have a high output.

Information about using open-collector outputs is contained in the Appendix. See Figure A-4.

The clean output, reliability, small size, and cheapness of Hall sensors have encouraged their use in devices as different such as hard drives, cameras, keyboards, and automobiles. They are useful in almost any situation where a sensor must detect a mechanical operation at close range. Three through-hole Hall sensors are shown in Figure 3-15. Surface-mount versions are much smaller.

Figure 3-15 *Three through-hole Hall-effect sensors. The background grid is in millimeters.*

How a Hall-Effect Sensor Works

When current is flowing through the length of a conductor, and a magnetic field is applied across the width of the conductor, the field generates a force causing electrons and electron holes to accumulate asymmetrically on opposite sides. This is known as the *Hall effect*.

The voltage difference between the electron-rich and electron-depleted zones is known as the *Hall voltage*. It is proportional to the magnetic field, and inversely proportional to the density of free electrons in the material. For this reason, the Hall effect is easiest to observe in semiconductors, where the density of electrons or electron holes is low.

Hall sensor components contain amplifier circuitry in addition to the sensing element itself. Typically there is an open-collector output, and a comparator or Schmitt trigger to provide some hysteresis.

Hall-Effect Sensor Variants

Four primary variants of Hall sensors are widely used.

Unipolar Hall Sensor

This is activated when an external magnetic field exceeds a threshold value. When the field diminishes, the sensor switches off. Unipolar sensors are available in versions activated by the north magnetic pole or south magnetic pole.

Bipolar Hall Sensor

Proximity to one magnetic pole will switch it on. Proximity to the opposite magnetic pole will switch it off. The sensor remains in its current state (on or off) in the absence of a magnetic field.

Omnipolar Hall Sensor

Proximity to a strong magnetic field of either polarity will switch it on. Removal of the magnetic field will switch it off. An omnipolar sensor can be thought of as a pair of unipolar sensors mounted in opposite directions and with their (open-collector) outputs wired together. This component functions similarly to a reed switch, although of course it still requires a power supply.

Linear Hall Sensor

Also known as an *analog* Hall sensor, its output voltage varies in proportion to an external magnetic field instead of switching cleanly between high and low states. When no magnetic field is detected, the output is half of the sensor's supply voltage. In response to one magnetic polarity, the output can drop almost to zero. The opposite polarity can increase the output almost to supply voltage.

The output from a linear sensor usually is supplied from the emitter of an internal NPN transistor, not the collector. A minimum 2.2K resistor should be connected between output and ground.

The variable output can be interpreted as a measurement of distance between the sensor and a magnet. In this mode, a Hall sensor functions as a **proximity** sensor. However, it is not usually capable of measuring a distance of more than 10mm.

Other Applications

Hall sensors are incorporated in other types of components. A **magnetometer**, for example, may contain Hall sensors.

Additional discussion of Hall-effect sensors, with test circuits, will be found in the book *Make: More Electronics*, from which some of the figures here have been excerpted.

Values

Magnetic field at operating point is the minimum field necessary for the output to switch on. It is measured in tesla or gauss, and the abbreviation B_{OP} is used.

Magnetic field at release point is the maximum field that allows the output to switch off. It is measured in tesla or gauss, and the abbreviation B_{RP} is used.

Magnetic field range may be specified for linear (analog) Hall sensors.

Supply voltage may range as widely as 3VDC to 20VDC, or may be restricted between 3VDC to 5.5VDC. Check datasheets carefully.

Sourcing or sinking capability for the open-collector output is typically 20mA.

How to Use a Hall-Effect Sensor

Hall sensors are often made in 3-pin packages. Through-hole variants are usually made of black plastic and look like TO-92 transistors, but are slightly smaller.

Surface-mounted variants are common.

A typical through-hole Hall sensor has one bevelled face and a flat face on the opposite side. The bevelled face may be referred to as the "front" of the component in a datasheet. The sensor responds when an appropriate magnetic pole is brought close to the front face of the sensor.

The part number printed on the front of the sensor may be abbreviated as three digits. The code below this usually refers to the date of manufacture.

A simple circuit for a Hall sensor resembles a typical circuit for a phototransistor. Positive supply voltage and negative ground are applied to two of the three leads. Positive voltage is also applied, through a pullup resistor, to the third lead, which is the open-collector output (except in the case of a linear Hall sensor, described previously). The output pin is then tapped as the output from the sensor, to be applied to a component that will not draw more than 20mA.

Configuration of Object Presence Sensors

While most of the following suggestions relate specifically to Hall-effect sensors, some general principles may be applied to optical sensors.

Linear Motion

A presence sensor can be activated when the triggering source (such as light or a magnet) approaches it directly. This is sometimes referred to as *head-on mode*. Alternatively, triggering can be arranged when the source moves past the sensor. This is sometimes referred to as *slide-by mode*. The two modes are illustrated in sections 1 and 2 of Figure 3-16.

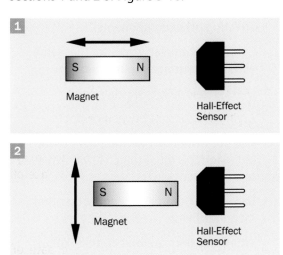

Figure 3-16 *Section 1 of this figure illustrates a presence sensor being used in head-on mode, while section 2 illustrates slide-by mode.*

Slide-by mode may be preferred because it eliminates the risk of damage to the sensor if overshoot occurs in head-on mode.

In slide-by mode, using a bipolar Hall-effect sensor, two magnets can be placed together with opposite polarity, creating a very steep transition in the overall magnetic field. This minimizes the risk of imprecise triggering. Using neodymium magnets, the triggering point can be adjusted with a precision of 0.01mm or better. See Figure 3-17.

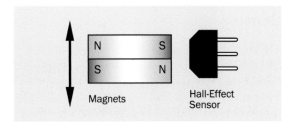

Figure 3-17 *In slide-by mode, two magnets with opposite polarity can be put together to create a very precise transition in a bipolar Hall-effect sensor.*

Sensing by Interruption

An *optointerrupter* is sensitive to an object passing between the light source and the light sensor. A comparable arrangement can be used with a Hall-effect sensor or a reed switch, but only if the interrupting object is thin and ferrous. This configuration is known as a *ferrous vane interruptor*.

Note that the magnet will exert significant force on the ferrous vane. This becomes an issue in sensors where the mechanical force is limited—for example, in paper-path sensors in a photocopy machine.

Additional information about the detection and measurement of moving objects will be found in the entry describing **linear position** sensors. See Chapter 6.

Angular Motion

One or more magnets can be used with a Hall-effect sensor to detect the angular motion, relative angular position, or absolute angular position of a rotating part. This data can be used to determine its speed of rotation. Some techniques to achieve this are described in the entry describing **rotary position** sensors. See Chapter 7.

Sensor Comparisons

Advantages of Optical Presence Sensors

- Not affected significantly by magnetic fields that can interfere with operation of a Hall-effect sensor or a reed switch.

- May be contained in a small, all-in-one package.

- Some optical sensors can operate over a distance of 50cm.

- Very well suited to sensing an object that blocks the light source (the *optointerrupter* configuration).

Disadvantages of Optical Presence Sensors

- Must have clear line-of-sight with the object and/or a reflector.

- Performance will be degraded by accumulation of dust or dirt.

- Limited lifetime of LED light source, if used continuously.

- May be accidentally triggered or impaired by some types of ambient light.

- Often requires a load resistor for the LED as well as a pullup resistor for the open-collector output.

- Range of acceptable voltage is usually narrow.

Advantages of a Reed Switch

- No polarity.

- No additional components required, other than a magnet.

- Able to switch AC or DC.

- Able to switch to voltages as high as 200V in some cases.

- Can be maintained by a magnet in open or closed state indefinitely without any power consumption.

- Many variants are capable of switching 500mA, and higher-current types are available.

- Can be activated through nonmagnetic materials (plastic, paper).

- Not significantly affected by dust or dirt that can degrade performance of an optical switch.

Disadvantages of a Reed Switch

- Requires a separate magnet (which must be placed carefully to avoid affecting other components).

- Cannot be miniaturized to the same extent as a surface-mount chip.

- Glass envelope is easily damaged.

- Arcing may occur between contacts.

- Will not work reliably when the activating magnet is more than a few millimeters from the switch.

- Can be activated accidentally by other magnetic fields.

- When sensing an object that comes between the switch and a magnet, only a ferrous object can be used.

- Must be debounced when connected with a logic chip or controller.

Advantages of a Hall Effect Sensor

- Robust solid-state component.

- Can be miniaturized for surface-mount applications.

- Very low cost.

- Fast response.

- No contact bounce.

- Extremely durable, with an almost unlimited lifetime.

- Not significantly affected by dust or dirt that can degrade performance of an optical switch.

- Low cost.

Disadvantages of a Hall Effect Sensor

- Requires a separate magnet (which must be placed carefully to avoid affecting other components).

- Open-collector output typically limited to 20mA or less.

- May be vulnerable to magnetic fields.

- When sensing an object that comes between the switch and a magnet, only a ferrous object can be used.

What Can Go Wrong

Optical Sensor Issues

Deterioration of LEDs

Most object presence sensors depend on infrared LEDs as a light source. An LED has many virtues (discussed in Volume 2), but does tend to suffer from gradual reduction in light output over a period of years. In a device such as a photocopy machine, which is used intermittently and may have an "economy" or "sleep" mode in which most of its components are powered down, LED-based detectors should last almost indefinitely. In other applications where an LED is powered continuously, its light output will diminish signicantly in 3 to 5 years. Bearing this in mind, an optical sensor should be chosen to operate well below its light-detection limits.

Object Too Close

Some optical and ultrasonic detectors triangulate on an object; that is, the light emitter and light sensor are angled slightly toward each other (as shown in sections 3 and 4 of Figure 3-3). The signal output from the sensor will peak at the distance where the emitter and sensor focus at a point. Consequently the output voltage will diminish when the object comes closer, which may create the misleading impression that the object is moving further away. To avoid spurious readings, detectors should not be used with objects that are closer than the minimum distance specified by the manufacturer.

Reed Switch Issues

Mechanical Damage

Bending the axial leads on a reed switch can easily fracture the glass envelope. Reed switches must be handled with care.

Contact Bounce

If the switch is wired to the input of a logic chip or microcontroller, contact bounce when the switch opens or closes is likely to be misinterpreted as multiple switching events. Debouncing will be necessary, either by additional components or by insertion of a momentary delay in the code embedded in a microcontroller.

Arcing

When switching high voltages or currents, an arc may be created briefly between the switch contacts, most often when they are moving from the closed to the open state. The arc erodes the contacts of the switch. Inductive loads make the arcing problem worse. If the switched voltage is kept below 5V arcing generally does not occur, extending the life of the switch.

passive infrared sensor

4.

In coloquial speech, the term *motion sensor* is usually understood to mean a **passive infrared** motion sensor.

The acronym *PIR* is often used for a passive infrared sensor. It is always capitalized, without periods.

Object presence sensors and **proximity** sensors require an active source of a magnetic field, ultrasound, or infrared radiation. A passive infrared sensor does not require any such source, and responds passively to heat radiated from the object that is being detected.

OTHER RELATED COMPONENTS

- **object presence** sensor (see Chapter 3)

- **proximity** sensor (see Chapter 5)

What It Does

A **passive infrared** motion sensor, often described as a *PIR*, detects *black-body radiation*, which all objects emit as a function of their temperature relative to absolute zero. The sensor responds to infrared radiation centered around a wavelength of 10µm (10 microns, or 10,000nm). This is the approximate body temperature of people and animals.

The word "passive" in the term "passive infrared" refers to the behavior of the detector, which receives infrared radiation passively. **Proximity** sensors must generate their own infrared radiation actively, which is interrupted or reflected by nearby objects. See Chapter 5.

Schematic Symbols
Schematic symbols that are sometimes used to represent a passive infrared motion sensor are shown in Figure 4-1.

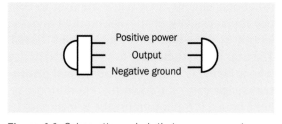

Figure 4-1 *Schematic symbols that may represent a passive infrared motion sensor. The orientation (pointing left or right) is arbitary. The pin sequence may vary.*

Applications
Motion-sensitive outdoor lighting almost always is based around a PIR. Similarly, a security system may sound an alarm or activate a video camera when a PIR indicates human activity.

Wildlife monitoring systems use PIRs to start a video camera that can then run for a preset interval.

Warning systems for automobiles have been developed that use a rear-facing PIR to detect pedestrians.

Industrial indoor lighting may use PIRs that switch the lights on automatically when people enter a room, and then switch the lights off (after a timed delay) when people are no longer detected in the room. The goal is to prevent wastage of energy as a result of employees forgetting to switch the lights off.

How It Works

A PIR module contains multiple components. Most visible is an array of at least 15 small lenses that focus infrared light from zones in the environment onto a *pyroelectric detector*, also known as a *pyroelectric sensor*. The response of the detector is processed by an amplifier, so that the signal can trigger an electromechanical **relay** or **solid-state relay** (see Volume 2). The relay operates an external device such as a light or an alarm.

Additional circuitry may allow the user to control the sensitivity of the PIR module and the length of time that the relay remains closed. The user may also be able to set the time of day when the PIR is active, or an additional **phototransistor** can shut down the PIR during daylight hours. If a phototransistor is included, its sensitivity is adjustable. For more information about phototransistors, see Chapter 22.

Pyroelectric Detector

The pyroelectric detector is actually a type of *piezoelectric* device. It is based around a wafer of lithium tantalate, which generates a small voltage in response to incident thermal radiation. However, like other piezoelectric components, it does not respond to a steady-state input, and must be activated by a transition. This distinguishes it from other types of light sensors, such as an infrared **photodiode**, in which the response is consistently related to a temperature input.

The response of a pyroelectric detector is suggested by the graphs in Figure 4-2.

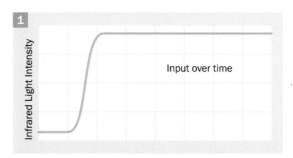

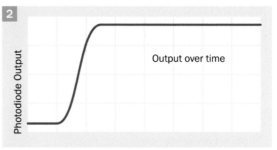

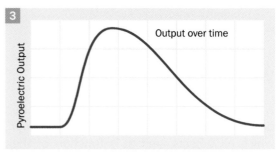

Figure 4-2 *Top: incident infrared light intensity. Center: the voltage output from a hypothetical photodiode. Bottom: the voltage output from a hypothetical pyroelectric detector.*

A pyroelectric detector in a PIR module is mounted in a sealed metallic container, as shown in Figure 4-3. The rectangular window in the detector is usually made of silicon, which is opaque to visible wavelengths but transparent to long-wave infrared radiation.

Elements

The pyroelectric detector used in a PIR contains at least two *elements* with opposite polarities, connected in series. If a sudden change in temperature affects both elements equally, their responses will cancel each other out. Thus, the detector ignores changes in ambient temperature. However, if a source of infrared radiation in the appropriate waveband affects one element before the other, the detector will emit two pulses of opposite polarity. See Figure 4-4.

Figure 4-3 *A pyroelectric detector mounted on a small circuit board in a passive infrared radiation sensor.*

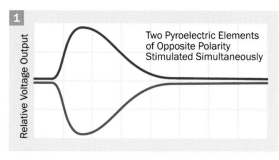

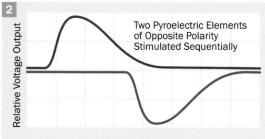

Figure 4-4 *Top: in a pyroelectric detector, if a change in temperature affects two elements of opposite polarity simultaneously, their voltages cancel each other out. Bottom: if one element is triggered before the other, the detector emits a signal.*

Lenses

Lenses are used to trigger the elements sequentially. Each lens faces toward one visible zone in the target area. When a source of infrared radiation moves from one zone to the next, it energizes the elements alternately, creating an output.

In some PIRs, four pyroelectric elements are used instead of two, to provide better coverage. The pairs of sensors may be wired in series or in parallel, but the principle remains the same.

The lenses are molded into a polyethylene dome that is often white and covers the pyroelectric detector. The dome appears smooth on the outside, but fine patterns of concentric ridges are molded on the inside. These are *fresnel lenses*, which are much cheaper, smaller, lighter, and easier to fabricate than conventional optical lenses. A fresnel lens introduces some distortion and aberration, but these defects are unimportant in a PIR.

Figure 4-5 illustrates the principle of a simple fresnel lens. The first section of this figure shows a conventional optical lens with one flat side and one curved side. A distant object emits almost-parallel beams of infrared light that are focused by the lens. Section 2 shows the same lens divided into segments that are stacked with no space between them. They behave in exactly the same way as the original lens. In section 3, each segment has been reduced in width, but because the optical faces still have the same geometry, they will still have the same function, although a small amount of distortion will be added by the reduction in width. This is a fresnel lens, which found an early application in lighthouses, where it greatly reduced the weight of very large glass lenses that focused the beam.

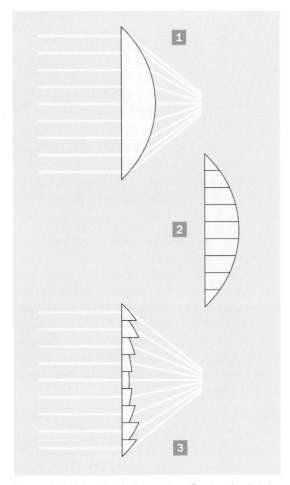

Figure 4-5 *Principle of a fresnel lens. See text for details.*

The same principle can be applied to a lens in which both surfaces are curved, as shown in Figure 4-6. In practice this will introduce more imperfections in the image, although some compensation is possible by adjusting the exact shape of the lens.

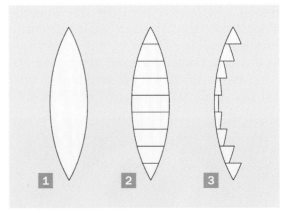

Figure 4-6 *The principle of a fresnel lens applied to a conventional lens in which both surfaces are curved.*

Figure 4-7 shows three curved fresnel lenses placed edge-to-edge, seen from above. In section 1 of this figure, infrared rays from a distant source are focused by the first lens onto the right-hand element of a pyroelectric sensor. In section 2, the external source has moved laterally, and the rays now focus on the left-hand element. In section 3, the source has moved into the zone covered by the centrally located fresnel lens, which focuses on the right-hand element again. The fluctuating inputs will trigger the sensor.

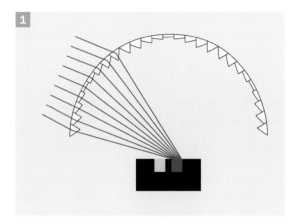

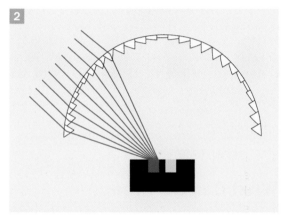

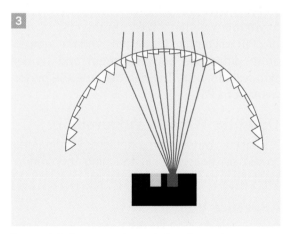

Figure 4-7 *Sections 1, 2, and 3 show fresnel lenses focusing an external source of infrared radiation on individual elements of a pyroelectric sensor.*

PIRs may combine fresnel lenses in a variety of patterns. Figure 4-8 shows an evenly weighted mosaic that would be suitable for a motion detector mounted on a ceiling, facing directly downward. Figure 4-9 shows a pattern weighted toward lateral motion, providing less sensitivity for motion above and below the primary band.

Figure 4-8 *An evenly weighted mosaic of fresnel lenses.*

Figure 4-9 *A mosaic weighted more toward sensing lateral motion. The grooves of the fresnel lenses are visible.*

Variants

PIR sensor modules are available mounted on a small board such as the one shown in Figure 4-10 from Parallax, Inc. The detection range is 5 to 10 meters, selected by a jumper on the board. The three pins visible in the photograph are for power supply (3VDC to 6VDC), ground, and output. The output can source up to 23mA with a 5VDC power supply. Power consumption of the module is only 130µA when it is idle, or 3mA when it is active but has no load.

Figure 4-10 *A passive infrared detector mounted with basic necessary components on a small board.*

A board of this type still requires additional components to set the "on" time for a light or alarm, and to deactivate the PIR during daylight hours.

Various lens patterns are available, sold separately.

A PIR can be bought as a single component containing two elements and FET transistors to amplify the signal. Surface-mount and through-hole versions are available, requiring a typical power supply of 3VDC to 15VDC.

However, a PIR bought as a "bare" component requires significant external circuitry, using comparators or op-amps. Circuit design is non-trivial, entailing practical problems such as op-amps being sensitive to voltage spikes caused by activating a relay that shares the same power supply.

An easier alternative is an all-in-one detector, lens, and control circuit such as the Panasonic AMN31111, which is ready for board mounting. Its small output current of 100µA would be capable of activating a solid-state relay. Similar Panasonic PIRs offer a variety of ranges, sensitivities, and supply voltages.

The AMN31111 is in Panasonic's AMN series. There are many type numbers for combinations of analog or digital output, lens shape, and black or white lens. A selection of lens shapes, derived from the manufacturer's datasheet, is shown in Figure 4-11.

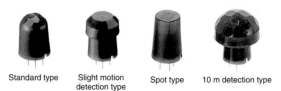

Standard type | Slight motion detection type | Spot type | 10 m detection type

Figure 4-11 *Four lenses from the Panasonic AMN series of passive infrared sensors.*

What Can Go Wrong

Temperature Sensitivity

In warmer weather, objects in the field of view of a PIR will tend to be warmer, and the temperature difference between them and human skin will diminish. This can degrade the performance of a PIR.

Detector Window Vulnerability

The silicon window on a detector is vulnerable to dirt or grease. Avoid touching the component if it is not protected by lenses.

Moisture Vulnerability

Water absorbs far-infrared light. Consequently, condensation on the lens or detector can degrade performance, and a PIR may not function in heavy rain or snow.

proximity sensor

Proximity sensors that use infrared, ultrasound, and capacitance are described here. This Encyclopedia does not include proximity sensors that use magnetism, inductance, or other methods of determining distance.

Sometimes proximity sensors are referred to as *distance sensors*. An ultrasonic proximity sensor may be described as a *range finder* or a *ranger*.

High-end ultrasound proximity sensors sold as large, sealed modules with cabling can monitor the status of industrial processes. These commercial units are outside the scope of the Encyclopedia.

A sensor that detects whether an object is present, but does not measure the distance to it, is considered to be an **object presence** sensor and has its own separate entry. See Chapter 3. Many devices that are sold as proximity sensors or distance sensors actually do not provide meaningful distance data, and therefore in this Encyclopedia they are included with presence sensors.

Phototransistors and **photodiodes** may be used as sensing elements in proximity sensors. These components have their own entries as light sensors. See Chapter 22 and Chapter 21.

OTHER RELATED COMPONENTS

- **object presence** sensor (see Chapter 3)
- **passive infrared** sensor (see Chapter 4)

What It Does

A proximity sensor measures the distance between itself and a physical object that is often described as the *target*. The output from the sensor may be analog (voltage), serial data, or pulse-width modulation. It may be transmitted via a serial protocol such as SPI, TTL, or I2C, and may be stored as digital data in a register that is accessed by a microcontroller, using I2C. For additional details about protocol, see Appendix A.

Schematic Symbols

Either of the schematic symbols in Figure 5-1 may represent a proximity sensor, but are not used consistently. The sensor may also be represented as a rectangle containing text describing its function.

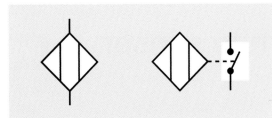

Figure 5-1 *Two ways to represent a proximity sensor in a schematic.*

Applications

In robotics, a proximity sensor can prevent a collision between a robot and an object or barrier in front of it. Some sophisticated proximity sensors can provide sufficient data for software to map the entire environment, but are outside the scope of this entry.

Proximity sensors can also be used in alarm systems, or for liquid-level sensing in storage tanks, or in automobiles to sound a warning if the driver is backing into an obstacle (although these devices are being augmented with rear-view video monitors).

In handheld devices, proximity sensors are used to sense the presence of the user's hand or face—for example, to shut down the display when a person raises a phone to talk into it.

Variants

This entry is subdivided to describe proximity sensors that use ultrasound, infrared light, and capacitance.

Ultrasound

An ultrasonic proximity sensor functions by emitting a short burst of sound and then listening for echoes from objects in front of it.

The sound is created by a **piezoelectric transducer** (see Volume 2) at a frequency between 30kHz and 50kHz, well above the range that can be detected by the human ear. The transducer may do dual duty as a microphone, sending and receiving sound on an alternating

basis; or a second transducer, serving as a microphone, may be mounted beside the emitter on a small circuit board. The economically priced HC-SR04 is an ultrasound proximity detector popular in the robotics community, working reliably in the range of 2cm to 5m.

A board on which the sensor is mounted may include a microcontroller to measure the delay between propagation of a pulse and reception of the echo. Distance to the reflecting object is then calculated using the speed of sound in air at sea level, which is approximately 340 meters per second.

Infrared

An infrared proximity sensor requires a beam of infrared light from an LED that may be incorporated in the sensing module or mounted separately. Light reflects from the target and is detected by a **phototransistor** or **photodiode**. From the angle of the reflected light, onboard electronics can calculate the distance to the target by a process known as *triangulation*. See Figure 5-2. (This diagram is simplified. An actual sensor may use a linear array of photodiodes to assess the angle of the returning light beam.)

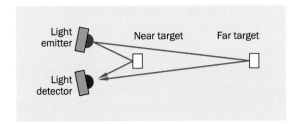

Figure 5-2 *An infrared proximity sensor determines the distance to an object by assessing the angle of reflected light.*

To reduce the risk of false positives, light from the LED contains only a narrow range of infrared wavelengths. Also, it is modulated at a frequency recognized by sensing circuitry in the module.

Relative Advantages

Ultrasound Devices

- Generally more suitable for detecting objects that are more than 1 meter away.

- Unaffected by direct sunlight, fluorescent tubes, and other light sources that can interfere with infrared devices.

- More accurate, often capable of placing objects within 5mm.

- Able to measure the distance to liquids and transparent objects, which may not be easily detected with infrared.

Infrared Devices

- Physically smaller—especially surface-mount versions.

- Able to measure the distance to soft objects, which may not be easily detected by ultrasound.

- More appropriate for targets that are closer than 10mm.

- More affordable.

Ultrasonic Examples

The proximity sensor in Figure 5-3 is a low-priced model, popular in the robotics community. Manufactured by MaxSonar, it uses a single element to send and receive. The manufacturer claims that it can detect a 6mm (quarter-inch) dowel 1.8 meters directly in front of it, and a 9cm dowel at 3.3 meters. This performance is beyond the capability of almost all infrared sensors.

MaxSonar offers a variety of sensors that appear physically very similar but have different range capabilities. Each has three outputs that can be accessed simultaneously: serial data at 9,600bps, analog voltage, and pulse-width modulation.

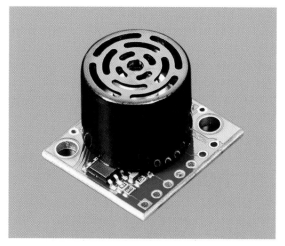

Figure 5-3 *The MaxSonar MB1003 can detect large solid objects as far away as 5 meters. Photo derived from an image by Adafruit.*

The serial output uses the RS232 protocol and consists of the letter R followed by four ASCII-coded numerals representing the measured range in millimeters. Thus, R1000 would indicate an object at 1 meter distance.

The analog voltage ranges linearly from 293mV when sensing an object at 300mm to 4,885mV for an object at 5,000mm.

The pulse-width output sends pulses ranging from 300μs indicating an object 300mm away to 5,000μs indicating an object 5 meters away.

The unit incorporates a temperature sensor that compensates for the lower density of air when its temperature increases. Weatherproofed versions are available. The power supply is 5VDC and must be smoothed.

Imports

Some international sources offer minimal paired ultrasound components at a very low price, such as the HC-SR04 from Cytron Technologies in Malaysia. See Figure 5-4.

Sound dispersion of the transducer is claimed to be plus-or-minus 15 degrees, and the claimed range is up to 4 meters. The module requires a triggering input pulse that must last at least 10μs. This prompts it to emit eight rapid ultrasound signals at 40kHz. The module measures the response time and applies a high state to its echo pin for a duration that is proportional with the distance measured. An external microcontroller must time the duration, then divide by a factor of 58 to obtain distance in centimeters.

Figure 5-4 *The HC-SRO4 is a very low-cost import that can provide acceptable performance when used with an external microcontroller.*

Many online sources offer simple code libraries for Arduino or PICAXE microcontrollers for use in conjunction with the HC-SR04.

Individual Elements

Individual ultrasonic components such as those shown in Figure 5-5 are available from many vendors. The user must add circuitry to generate a high frequency for a short duration, amplify the microphone signal, measure the time difference, and calculate distance.

Figure 5-5 *These components, listed as the "40TR12B-R ultrasonic sensor kit" by the online supplier Jameco, could form the basis of a DIY ultrasonic proximity sensor.*

Infrared Examples

Sharp manufactures four infrared proximity sensors that are widely regarded as accurate and easy to use. They are popular in the robotics community. Their part numbers and ranges are:

- GP2Y0A51SK0F (20mm to 150mm)
- GP2Y0A21YK0F (10cm to 80cm)
- GP2Y0A02YK0F (20cm to 150cm)
- GP2Y0A60SZLF (10cm to 150cm).

The GP2Y0A60SZLF is the most recent product, with the most impressive specification. The GP2Y0A21YK0F is shown in Figure 5-6.

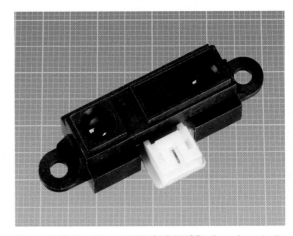

Figure 5-6 *The Sharp GP2Y0A21YK0F infrared proximity sensor. The background grid is in millimeters.*

Sharp describes these sensors as having an *analog output*. The voltage on an output pin varies in inverse proportion with the distance being measured. The relationship is illustrated by the graph in Figure 5-7, derived from the manufacturer's datasheet for model GP2Y0A02YK0F.

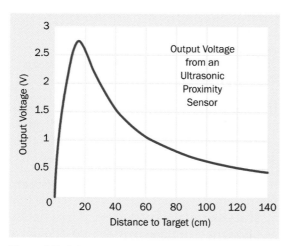

Figure 5-7 *Relationship between output voltage and target distance for a Sharp infrared proximity sensor, derived from the manufacturer's datasheet.*

The Sharp sensors can work with 5VDC. They consume around 30mA, except for the GP2Y0A60SZLF, which uses less current. Because the infrared LED functions in bursts, the manufacturer recommends protecting other components sharing the power supply by placing a 10µF capacitor across the sensor supply pins.

Trends in Infrared Proximity Sensing

Like many other types of sensors, proximity sensors have been affected by the huge market for handheld devices.

Handheld applications have had four consequences:

Miniaturization
> Infrared proximity sensors are now commonly found in surface-mount chips measuring 5mm × 3mm or smaller.

Onboard processing
> The status of a photodiode can be processed by a microcontroller on the same chip, to determine what the sensor is really "seeing." Input from an included ambient light sensor is factored into the evaluation.

Cost reduction
> While chip-based proximity sensors have become increasingly sophisticated, their unit cost has plummeted, so that they are now actually much cheaper than simpler devices such as the Sharp analog sensors described above.

Complexity
> Modern sensors must be programmed with a complex variety of instructions, and their coded output must be interpreted with a separate microcontroller running its own program. Whether this compensates for the low price and added functionality is something for the individual developer or experimenter to decide.

The Silicon Labs Si1145/46/47 series are chips with the kind of sophisticated capabilities required for handheld devices. An external microcontroller communicates with the sensor via an I2C connection, and can instruct it to adjust its distance range (from 1cm to more than 50cm), its analog-to-digital conversion sensitivity, and its current-sinking capability for up to 3 external LEDs. The chip incorporates ultraviolet sensing and ambient-light sensing capability. Its I2C connection can run at up to 3.4Mbps. Because its light output is pulsed for only 25.6µs every 800ms at 180mA, the average power consumption is only 9µA, assuming a supply voltage of 3.3VDC.

For additional details about protocols such as I2C, see Appendix A.

In addition to its use in handheld devices, the manufacturer suggests applications including heart-rate monitoring, pulse oximetry, and display backlighting control. These applications require only a subset of the features built into the sensor, but its price is so low, it may be cost-effective even when many of its capabilities are unused.

Proximity sensors with comparable specifications are available from many manufacturers. Examples are the Vishay VCNL4040 and the Avago HSDL-9100. Figure 5-8 shows the Silicon Labs SI1145 on the left, and the Avago HSDL-9100 on the right.

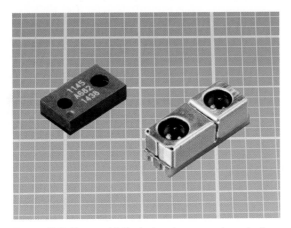

Figure 5-8 *Two sophisticated surface-mount proximity sensors with digital output. Left: Silicon Labs SI1145. Right: Avago HSDL-9100. The background grid is in millimeters.*

Several guidelines must be observed when using this kind of sensor. First and most obviously, if an external LED is used, it must have a peak wavelength compatible with the photodiode in the sensor. The LED must be placed as near as possible to the photodiode, as reducing the separation increases the sensitivity; but crosstalk between the photodiode and LED must be minimized, usually by placing a thin barrier between them that is barely taller than the higher component.

If the light emitter and/or light sensor are protected behind a transparent glass or plastic panel, it must have minimal resistance to infrared wavelengths, and its thickness must be chosen in compliance with guidelines in the sensor datasheet. To prevent light from reach-

ing the sensor by reflection from the rear of the panel, a thin, opaque tube can be installed between the LED and the panel.

A sensor of this type may be configurable with a "detection scheme," meaning high and low sensing threshold levels appropriate to the object that is likely to be detected. Determining this may be a process of trial and error.

Capacitive Displacement Sensor

This is also known as a *capacitive linear displacement sensor*. It should not be confused with a capacitive **single touch** sensor, a human-input device that has its own entry. See Chapter 13.

A **capacitive displacement** sensor measures the distance between itself and a target that must be electrically conductive. Unlike optical or ultrasound position sensors, no additional source of light, sound, or other radiation is required. Unlike a magnetic position sensor, it does not require a separate permanent magnet. It simply measures its electrical capacitance with the target.

High-precision capacitive displacement sensors are used mainly for industrial process management. Lower-precision variants are much less expensive and can be used as **object presence** sensors to determine if an object is anywhere within a specified range.

A typical maximum range would be 10mm. For larger distances, optical and ultrasound sensors are more appropriate.

A selection of cylindrical sensor probes is shown in Figure 5-9.

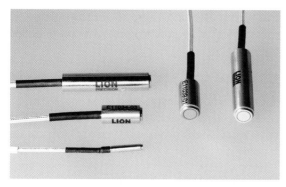

Figure 5-9 *Some high-precision capacitive displacement sensors from Lion, which builds them into cylindrical probes ranging from 3mm to 18mm in diameter.*

Applications

The high-precision variant of this type of sensor is commonly used during the production of small devices, such as disk drives. It can also measure vibration of a rotating metal part, such as a motor shaft, and may maintain automatic focus of a microscope.

Lower-precision variants can be used for applications such as counting objects on a conveyor.

When used to measure materials thickness, the sensor finds applications in checking automobile brake rotor fabrication and the thickness of silicon wafers.

How It Works

When two plates of electrically conductive material are placed opposite each other, they possess capacitance. This means that they have an electrical storage capacity enabled by accumulation of opposite charges on each plate.

The capacity is directly proportional to the plate area, inversely proportional to the distance between the plates, and is affected by the medium separating the plates, known as the *dielectric*.

If the plate area and dielectric remain constant, the distance between the plates will be the only factor affecting the capacitance. Therefore, the distance between the plates can be calculated by measuring the capacitance.

Measurement can be performed by evaluating the *displacement current* that passes through the dielectric from one plate to another when a pulse of voltage is applied. (Hence the term "capacitive displacement" in the name of this type of sensor.)

A detailed explanation of displacement current is included in the second edition of *Make: Electronics*.

The sensor itself functions as one plate of the capacitor while the target functions as the opposite plate. Alternating current is applied as a rapid series of pulses, and the current that passes between the plates is proportional to the distance between them.

Ideally, the target should be grounded to the current source. However, since AC is being used, capacitive coupling of the target to the current source is also possible, so long as the additional capacitor has a value of 0.1μF or higher.

Sources of Error

To obtain a meaningful measurement, the electric field from the sensor is focused on the target. Some dispersion will still occur, and the combination of sensor and target must be compatible. A target typically should be flat and should have a larger surface area than the sensor.

Humidity can affect sensor performance, as it changes the value of the dielectric. Temperature can affect performance, partly because it causes small dimensional variations in the sensor and the target.

The surface of the sensor and the surface of the target should be precisely parallel, because the spot where the field hits the target will be elongated if the target is tilted relative to the sensor. Elongation increases the capacitive area and affects the accuracy of measurement.

This type of sensor can also measure the thickness of a nonconductive material, if the material is a thin sheet that can be interposed between two sensors. In this mode the material functions as a dielectric, and its thickness will affect the AC current passing through it.

While lower-precision capacitive displacement sensors are relatively unusual, they can potentially serve as a relatively affordable and simple **object presence** sensor, so long as the target is conductive and will not be damaged by passing a small alternating current at a relatively low voltage.

Values

A high-precision capacitive displacement sensor can measure distances usually ranging from 0.25mm to 10mm with an accuracy that can be better than 0.05mm. High voltages are not required, with a supply of plus-or-minus 15V being common.

The sensing element, often referred to as a *probe*, is usually plugged into a custom-designed control unit that converts the capacitance measurement into a variable output voltage. The performance is then expressed in millimeters per volt. Thus if the voltage varies by 5V over a distance of 1mm, the sensor is rated as providing 0.2mm per volt.

What Can Go Wrong with Optical and Ultrasound Proximity Sensors

Object Too Close

Because both types of proximity sensors (ultrasonic and infrared) may include emitters that are angled toward the distance range for which they are designed, they may fail to "see" an object that is closer. Consequently the sensor will provide no response, or may detect a different object that is further away. In either case, if the sensor is being used on a moving device, the device may collide with the undetected near object.

Multiple Signals

If two or more sensors and emitters are used concurrently, their combined signals can interfere with each other and create inaccurate readings.

Inappropriate Surfaces

Ultrasonic proximity sensors are intended to identify a single object that is closer to the sensor than other objects, within a narrow beam dispersion angle. Multiple objects, complex surfaces, soft surfaces such as clothing or furnishings, or an unusual configuration of interior walls can create inaccurate readings.

Infrared sensors may be unable to "see" liquids or transparent objects, and may give different assessments of distance depending on the properties of a surface. Human skin, for example, is a poor reflector, as it absorbs some infrared radiation.

Environmental Factors

An ultrasonic transducer uses a very small moving diaphragm to generate sound. Like any system containing moving parts, it will be vulnerable to moisture or excessive humidity, and may need to be protected.

After a device is built and tested indoors, in a controlled environment, it is likely to behave differently if it is moved outside where the temperature is significantly higher or lower.

Deterioration of LEDs

As noted in the section on presence sensors (see Chapter 3), LEDs tend to suffer from gradual reduction in light output over time. The performance of an infrared proximity sensor may deteriorate over a period of years, depending on how intensively it is used.

linear position sensor

A **linear position** sensor may also be described as a *linear displacement sensor* or a *linear position transducer*. It is sometimes categorized as a type of **proximity** sensor, but in this Encyclopedia a proximity sensor emits a signal and receives an echo to measure the distance from itself to an object. By comparison, a linear position sensor measures the position of a sliding object within a stationary enclosure.

An **object presence** sensor can be considered as a form of linear position sensor, but it only responds to the presence of an object, without measuring its position.

OTHER RELATED COMPONENTS

- **proximity** sensor (see Chapter 5)

- **object presence** sensor (see Chapter 3)

- **rotary position** sensor (see Chapter 7)

What It Does

Control of a mechanical device may require accurate and timely information about the position of a movable part in the device. A **linear position** sensor can be used for this purpose.

Three attributes are likely to be of interest:

- Position

- Direction of motion

- Speed of motion

Typically a linear position sensor measures only the first attribute. Additional electronics may calculate the second and third attributes by taking multiple position readings. Thus a *speed sensor* is very likely to be built around a position sensor, and therefore this Encyclopedia does not have a separate entry for speed sensors.

Applications

Robotic arm positioning, wing flap and rudder positions on an aircraft, computer-controlled machine tools, 3D printers, and automobile seat position sensors are some of the many applications.

Schematic Symbol

In a schematic, a linear position sensor may be represented by symbols for the sensing elements that are inside it (potentiometer, LED, phototransistor, or others).

How It Works

Linear potentiometers, *magnetic linear encoders*, *optical linear encoders*, and *linear variable differential transformers (LVDTs)* are described here. Other options are available, but they tend to be more specialized, and are not included.

Linear Potentiometer

For a full description of **potentiometers**, see Volume 1.

A linear potentiometer, often referred to as a *slider potentiometer*, contains an electrical resistance in the form of a straight section of *track*. The track may be a strip of *resistive polymer* or (less often) may consist of an insulator with a coil of nichrome wire wrapped around it.

For sensing purposes, the potentiometer is wired as a voltage divider, and a fixed potential is applied across its full length, as shown in Figure 6-1. A *wiper* slides along the track, sensing a voltage that varies linearly with the wiper's position. Output from the wiper can be used directly to control an analog indicator such as a meter, or can be processed by an analog-to-digital converter.

Figure 6-2 *A linear potentiometer contained in a tube with sealed bearings.*

Small linear potentiometers for position sensing are available from companies such as Bourns. The one shown in Figure 6-3 is about 20mm long, and the rod that slides through it has a travel of approximately 10mm. The component is available with resistance values ranging from 1K to 50K, and its power rating is 1/8 watt. The manufacturer claims a life expectancy of 500,000 cycles.

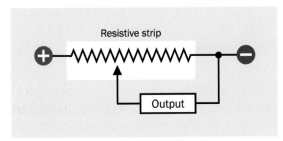

Figure 6-1 *A linear potentiometer can consist of a fixed track with a known voltage applied at each end, and a movable wiper.*

For audio applications, a slider potentiometer may have resistance that varies logarithmically with the position of the wiper. However, this type of component is not generally used as a position sensor.

For sensing purposes, the potentiometer is usually protected by a long, narrow box or tube through which a rod slides on sealed bearings. An example is shown in Figure 6-2.

Figure 6-3 *A miniature linear potentiometer. The body is about 20mm long.*

A linear potentiometer is simple, inexpensive, compact, and requires few additional components. The track contains a lubricant, but inevitably some wear occurs as a result of motion of the wiper. Life expectancy will be reduced by vibration or by contamination with dirt or moisture.

A linear potentiometer may rarely be described as a *linear potentiometric sensor*.

Magnetic Linear Encoders

A ferrous rod or strip can be magnetized with alternating north and south poles. When it slides past a single bipolar Hall-effect sensor, it generates a pulse train from the sensor that can be interpreted to provide positional information. The principle of the component is shown in Figure 6-4. (Magnetic rotary encoders also exist; see "Rotary Encoders".)

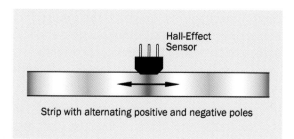

Figure 6-4 *When a strip magnetized with alternating north and south poles slides past a sensor, the pulse train from the sensor can be decoded to indicate the relative position of the strip.*

The sensing element may be described as a *read head*. If two are used, with a spacing equal to half the interval between the north and south poles on the strip, the phase difference between the pulse trains from the sensors will indicate the direction in which the magnetized strip is moving. This is shown in Figure 6-5.

The combination of pulse trains is known as *quadrature*, because there are four possible combinations: A and B both high, A and B both low, A high and B low, or A low and B high. The same principle is used in an optical rotary position sensor; see Figure 7-6.

This type of linear position sensor is often described as a *magnetic encoder*, meaning that the position of the sliding part is encoded in a series of pulses. Relatively high resolution is possible, as the north and south poles on a strip of ferrous material can be as close as 2mm. Optical encoders may use the same principle; see "Optical Linear Encoders".

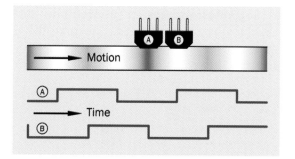

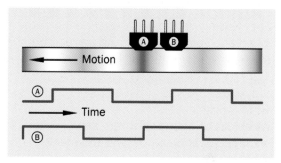

Figure 6-5 *Pulse trains from Hall-effect sensors A and B are shown in green. The phase difference between them can be interpreted to show the direction in which the magnetized strip is moving.*

The sensor can be built into a module containing an analog-to-digital converter that provides numeric output defining the position of the read head.

In an *absolute* magnetic encoder, nonvolatile memory can store the digitized position when the device is switched off. An *incremental* magnetic encoder does not store this information, and will require at least one additional *home sensor* to detect when the encoder is at either end of its travel. At power-up, an initialization routine moves the magnetized strip until the home sensor is triggered.

For details of Hall-effect sensors see "Hall-Effect Sensor".

Optical Linear Encoders

The operation of an optical linear encoder is identical to that of the magnetic linear encoder described immediately above, except that a sliding *optical grating* is used in conjunction with a light source and a detection device such as a **phototransistor** or **photodiode** that functions as the read head. The principle is shown in Figure 6-6. The grating may be described as a *codestrip*.

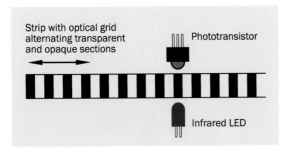

Figure 6-6 *An optical linear encoder uses the same general principle as a magnetic linear encoder.*

For details of phototransistors, see Chapter 22. For details of photodiodes, see Chapter 21.

An example of a low-cost optical linear encoder is the Avago HEDS-9 series, which consists of a horseshoe-shaped module with an LED in one arm and an array of photodiodes in the opposite arm. When a codestrip passes between the arms, the module emits two pulse trains from internal comparators. The pulse trains are 90 degrees out of phase, and can be interpreted to show which way the codestrip is moving.

These sensors have a body measuring approximately 10mm wide and are designed to read opaque/transparent intervals ranging from 1.5 to 7.87 cycles per millimeter. The output rate can be as high as 20kHz. No pullup resistor is required on the output, as a 2.5K resistor is integrated.

A *codewheel* may be used, in which case the module senses rotation instead of linear motion. This is described in detail in the entry on **rotary position** sensors. See Chapter 7.

Linear Encoder Applications

Optical or magnetic linear encoders are found in some laboratory equipment, machine tools, and industrial robots. The mean time between failures can range from 100,000 to 1,000,000

hours. Optical encoders must be sealed to provide good protection from dust and dirt.

Linear Variable Differential Transformers

This type of sensor, often referred to by the acronym LVDT, tends to be used in industrial environments where great reliability is required under severe conditions. Examples are high-temperature steam valves and nuclear reactor control mechanisms. However, the robust, frictionless design suggests other possible applications, and custom fabrication is a possibility.

Figure 6-7 illustrates the general principle. Three coils are wound sequentially around a (nonmagnetic) stainless-steel tube, enclosed in a second tube also made of stainless steel. The coils act as transformers with the variable voltage ratio determined by the position of a solid iron armature that slides through them.

The center coil is the primary winding, to which AC is applied between 2kHz to 50kHz, depending on the application. (The frequency must be at least ten times the maximum rate of motion of the armature.) The iron armature is attached to a nonmagnetic rod. The voltages on the secondary windings provide the output from the sensor.

While more than one wiring arrangement is possible, the most common schematic is shown in Figure 6-8. The secondary coils are in series, with one of them reversed, so that the phase of the output is inverted as the armature moves from one end of its travel to the other. The phase detector responds to the phase difference by creating a DC output that varies with movement of the armature. All the functions shown in the schematic are available on a single integrated circuit chip.

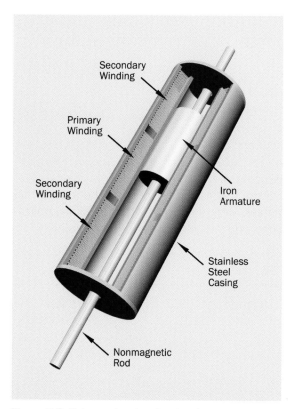

An example of a linear variable differential transformer is shown in Figure 6-9.

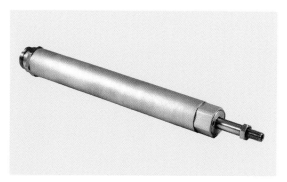

Figure 6-9 *External view of a linear variable differential transformer.*

What Can Go Wrong

Mechanical Issues

Any sliding mechanism will involve friction, and will be vulnerable to wear and tear, resulting in looseness that will degrade its accuracy. In addition, optical systems are vulnerable to dust and dirt.

LED Longevity

Light output of an LED diminishes over a period of years if the LED is "always on." This will limit the life of the sensor.

Figure 6-7 *Cutaway drawing showing the internal design of a linear variable differential transformer, in which the position of a sliding iron armature determines the voltage induced in the secondary windings.*

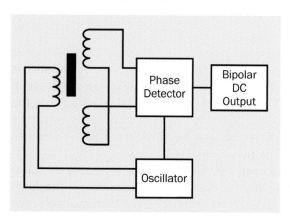

Figure 6-8 *Typical schematic for using a linear variable differential transformer.*

rotary position sensor

7

Alternative terms for **rotary position** sensor are *rotary sensor*, *rotational position sensor*, *angular position sensor*, and *angle sensor*.

The terms *rotary* and *rotational* are used interchangeably. In this entry, an attempt has been made to use the term that is most common for each specific application. For example, "rotary position sensor" versus "rotational encoder."

A few sensors are available that specifically measure *rotary speed*, but generally a rotary speed sensor uses information from a rotary position sensor. Therefore, this Encyclopedia does not include a separate entry for rotary speed sensing.

A **rotational encoder** can be used as a rotary position sensor. It is mentioned briefly here but is described in more detail in Volume 1 of this Encyclopedia in the entry describing **switches**.

OTHER RELATED COMPONENTS

- **linear position** sensor (see Chapter 6)

What It Does

Control of a mechanical device may require accurate and timely information about the orientation of a rotating part in the device. A **rotary position** sensor can be used for this purpose.

A sensor may be capable of measuring three attributes:

- Angular orientation
- Direction of rotation
- Speed of rotation

Typically a rotary position sensor measures only the first attribute. Additional electronics are required to calculate the second and third attributes by taking multiple position readings. Thus a *speed sensor* is very likely to be built

around a position sensor, and therefore, this Encyclopedia does not have a separate entry for speed sensors.

Applications

In robotics, a rotary position sensor is commonly used to show the orientation of a pivoting arm or strut. It can also be used as a limit switch on a motor.

Specific applications include solar array positioning, remotely piloted vehicles, guidance and navigation, antenna positioning, and wind turbine pitch control.

Pulses from a rotary position sensor are used to measure speed of rotation in vehicles, industrial processes, and aviation. Small rotary speed sensors are built into devices such as cooling fans and computer hard drives.

Schematic Symbol

In a schematic, a rotary position sensor may be represented by symbols for the sensing elements that are inside it (potentiometer, LED, phototransistor, and others).

Potentiometers

Single-turn and multiturn potentiometers may be used as rotary position sensors. For basic information about **potentiometers**, see the entry describing this component in Volume 1.

Arc-Segment Rotary Potentiometer

An *arc-segment rotary potentiometer* is often referred to simply as a "potentiometer," as this type is more common than the multiturn type or the linear type. When used as a sensor, it can measure a turn angle that is less than 360 degrees.

This component contains a resistor in the shape of an arc, referred to as the *track.* It may be a strip of *resistive polymer* or (less often) may consist of an insulator with a coil of nichrome wire wrapped around it.

For sensing purposes, the potentiometer is wired as a voltage divider, and a fixed potential is applied along the full length of the track, as shown in Figure 7-1. A *wiper* slides along the track, sensing a voltage that varies linearly with the wiper's angular position. Output from the wiper can be used directly to control an analog indicator such as a meter, or can be processed by an analog-to-digital converter.

For audio applications, an arc-segment potentiometer may have resistance that varies logarithmically with the position of the wiper. However, this type of component is not generally used as a position sensor.

Low-cost potentiometers were traditionally used as volume or tone controls in stereo systems. When a potentiometer is designed for use as a sensor, it tends to be more ruggedly

built and better protected against dust, dirt, and moisture. Its advantages are that it is simple, inexpensive, compact, and requires few additional components.

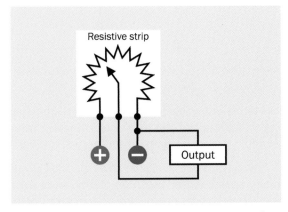

Figure 7-1 *An arc-segment potentiometer consists of a fixed track in the shape of an arc with a known voltage applied at each end, and a movable wiper.*

The major disadvantage is that although the track contains a lubricant, some wear will gradually result from friction with the wiper. Life expectancy will be further reduced if there is vibration or contamination with dirt or moisture.

End Stops

An arc-segment potentiometer usually has *end stops* to prevent the wiper from running off either end of the track. Typically, these stops limit rotation to around 300 degrees.

A few arc-segment potentiometers allow unrestricted rotation. The Bourns 6639 series is an example, although the wiper still passes through a "dead zone" of 20 degrees between the start and end of its track. An application for this type of potentiometer could be as a direction sensor for a weather vane.

Multiturn Rotary Potentiometer

A spiral track, which resembles a coil spring, enables a rotary potentiometer to make multiple turns. The wiper rotates inside the track and follows its contour. A potentiometer of this type will still be calibrated in degrees; thus, a 10-turn component may be listed as allowing an *electrical travel* of approximately 3,600 degrees.

The exterior of a multiturn rotary potentiometer is shown in Figure 7-2.

Figure 7-2 *A multiturn rotary potentiometer. Each pair of solder tags is attached internally to one end of the internal coiled track. The smaller tag at the far end connects with the wiper.*

A simpler type of multiturn potentiometer is intended for use as a trimmer—a small potentiometer that can be mounted on a circuit board to allow adjustment or calibration, often during the manufacturing process. This type of trimmer contains a worm gear that engages with a spur gear internally. The wiper is mounted on the spur gear. It has no applications as a sensor, but is mentioned here to avoid ambiguity, as it is probably the component that is most commonly referred to as a "multiturn potentiometer."

Magnetic Rotary Position Sensor

Externally, a modern magnetic rotary position sensor may look very much like an arc-segment rotary potentiometer. Internally, a permanent magnet is attached to the base of the shaft, and one or more **Hall-effect** sensors are mounted on a small circuit board immediately below the magnet, in the bottom of the enclosure. A simplified diagram appears in Figure 7-3.

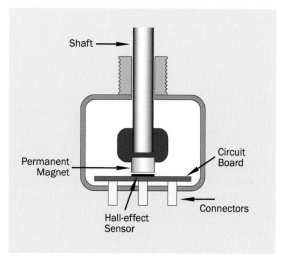

Figure 7-3 *Simplified interior view of a magnetic rotary position sensor.*

For more information about Hall-effect sensors, see "Hall-Effect Sensor".

A magnetic rotary position sensor may be described as a *noncontacting* sensor. Two views of an example are shown in Figure 7-4.

Figure 7-4 *Two views of a Bourns AMS22 magnetic rotary position sensor.*

The AMS22 sensor shown in Figure 7-4 has an analog output ranging from 0.1VDC to 4.9VDC when powered with 5VDC. A noncontacting sensor of this type will cost three to four times as much as a conventional potentiometer, but has the great advantage of extreme durability, with the manufacturer claiming a life of 50 million shaft rotations. A disadvantage is its low-current output, limited to 10mA. The maximum rotation speed of its shaft is 120rpm.

Rotary Position Sensing Chips

Chips of the type used in a magnetic rotary position sensor are available as individual components, many of which have advanced features. For example, the AM8192B angular magnetic encoder by RLS is a 44-pin surface-mount chip containing Hall sensors that detect the orientation of a permanent magnet above or below the chip. Various outputs provide information such as sine or cosine of the turn angle, incremental pulses, and digitized output via an SPI interface.

Rotary Encoders

Many rotating position sensors communicate their angle of rotation with an output that consists of a pulse train or some other coded signal. This type of rotary sensor is known as a *rotary encoder* or *rotational encoder*. (Linear encoders also exist; see "Magnetic Linear Encoders".)

The simplest version of this component is a *mechanical encoder* containing two small electromechanical switches that are activated, out of phase, by a toothed wheel attached to the rotating shaft. This component is described in detail in Volume 1 of the Encyclopedia, where it is categorized as being a form of **switch**. Its low cost and simplicity has made it popular for rotary controls on car radios and small stereo systems, but its switches have a limited life expectancy and create a "noisy" output that must be *debounced* if connected with a logic chip or microcontroller. Typically the microcontroller will include a pause of up to 50ms in its program code to allow time for the switch contacts to settle (although some manufacturers claim only 5ms).

Confusingly, a mechanical encoder is often identified only as a "rotational encoder," even though optical and magnetic rotational encoders also exist, as described immediately below. As a general rule, if a component is described simply as a rotational encoder, it probably contains electromechanical switches.

Optical Rotary Encoders

This type of component works on the same principle as an *optical linear encoder* of the type described in "Optical Linear Encoders". The difference is that a *codewheel* is used instead of a *codestrip*. Typically the codewheel is supplied by the manufacturer of the component that is designed to read it.

A *transmissive* codewheel is shown in Figure 7-5. The distance between the light emitter and the light detector has been exaggerated in this diagram for clarity.

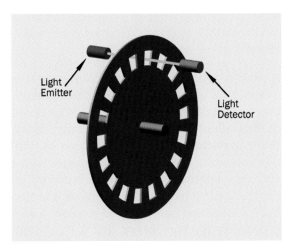

Figure 7-5 *A light-transmissive codewheel for use in an optical rotary encoder.*

Some optical rotary encoders use a *reflective* codewheel, in which case alternating sections of the wheel are light-absorptive and light-reflective, and the emitter and detector are both on the same side of the wheel.

If only one light emitter and light detector are used, the pulse train from the sensor reveals how many increments the wheel has turned relative to its previous position.

A single sensor cannot indicate the direction of rotation, but if a second emitter-sensor pair is added, 90 degrees out of phase with the first, a microcontroller can assess the phase difference between pulse streams to determine which way the wheel is turning. This principle is illustrated in Figure 7-6 where one transmitter-detector pair is located in position A and another is in position B. The resulting pulse trains, for clockwise and counterclockwise rotation of the wheel, are shown in green. The combination of pulse trains is known as *quadrature*, because there are four possible combinations: A and B both high, A and B both low, A high and B low, or A low and B high.

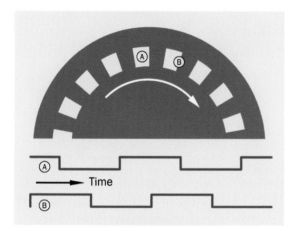

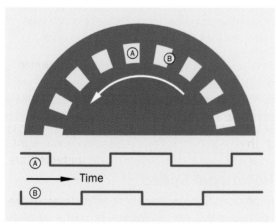

Figure 7-6 *With transmitter-detector pairs mounted at positions A and B, the phase difference between their pulse trains can show the direction of rotation of the wheel.*

This is the same principle as illustrated for a magnetic linear position sensor in Figure 6-5.

An alternative system of quadrature for an optical rotary encoder uses two separate tracks on the disc, each having an equal number of opaque and transparent sections, but half an interval out of phase.

Any system that reveals the relative motion of the wheel, but cannot determine its absolute angular position, is an *incremental sensor*.

Optical Products

High-end optical rotary encoders may use discs with as many as 600 sequential opaque and transparent segments to provide extremely high resolution. They are outside the scope of this Encyclopedia.

Moderately priced optical encoders are available as shaft-driven assemblies very similar in external appearance to potentiometers. Typically they have four terminals, one pair for power and ground connections and another pair for quadrature output from the two internal sensors, usually identified as A and B in datasheets. Some encoders also contain an on-off switch that is activated by pressing the shaft, in which case two additional terminals will be provided.

Bourns is a leading manufacturer of this type of encoder, an example being the EM14, which is mounted in a box-shaped body measuring 14mm square. It uses a 5VDC power supply and provides pulses of 4VDC minimum with intervals of 0.8VDC maximum. Variants are available with 8 to 64 pulses per revolution. Intended for audio applications, this type of encoder has a maximum rotation speed of 120rpm.

An example of an optical rotary encoder from a German manufacturer is shown in Figure 7-7. Its resolution is 25 pulses per rotation. The rectangular package measures approximately 19mm × 25mm. The power supply can be 3.3VDC or 5VDC.

Optical rotary encoders of this type cost about five times as much as mechanical rotary encoders at the time of writing, but their longevity and their clean output signals make them an attractive alternative, and the price difference may diminish over time.

Figure 7-7 *A compact incremental optical rotary encoder in the MRB25 series from Megatron Elektronik AG & Co.*

Encoders such as the Avago HEDS-9 series are not protectively enclosed, and require assembly with a codewheel supplied by the manufacturer. See Figure 6-6 for additional information.

A more basic optical rotary encoder, shown in Figure 7-8, is sold by Cytron Technologies as an *optical switch* mounted on a small board, with a separate codewheel consisting of a slotted disc. This low-cost kit is intended for use in DIY robotics. More information about optical switches will be found in the entry dealing with **optical presence** sensors. See Figure 3-3.

Figure 7-8 *A bare-bones optical encoder for DIY robotics.*

Computer Mouse Principles

The original design of a computer mouse, with a hard rubber ball, contained two optical rotary encoders oriented at right angles to each other. Each of them used a transmissive codewheel. The rolling ball turned the codewheels as the mouse was moved across a desktop, and electronics in the mouse converted the outputs from the encoders into a pulse train that could be interpreted by a computer. Figure 7-9 shows the primary components.

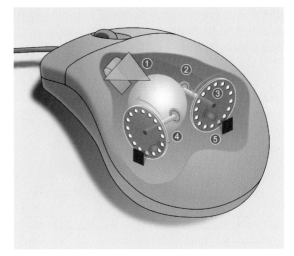

Figure 7-9 *From Wikimedia Commons, this rendering by Jeremy Kemp shows (1) rotation of the ball, (2) a roller touching the ball, (3) a transmissive optical codewheel, (4) an infrared LED that shines through a second codewheel, and (5) a sensor detecting the pulses of light.*

An *optical mouse* works on a different principle, maintaining a monochrome image of the desk surface on an optical array that functions like a very low-resolution camera sensor. Electronics in the mouse detect displacement of the image as the mouse is moved.

Rotational Speed

Incremental rotary encoders, which only supply relative data, are adequate in many applications, especially speed measurement, where a microcontroller can compare a pulse stream from a sensor with a target frequency, and can also provide feedback to control motor speed appropriately. This is convenient when a stepper motor is being used, or a DC motor that is controlled with pulse-width modulation. (See the entry on **motors** in Volume 1.)

Toothed wheels or magnetized wheels are commonly used to measure speed of rotation in applications ranging from automobile transmissions to computer disk drives.

While a second sensor can be added to determine the direction of rotation, this still does not provide information about the absolute orientation of a part.

Absolute Position

If associated electronics are equipped with nonvolatile memory, it may be used to store a sensor-wheel position from one session to the next. This may be sufficient in noncritical applications such as volume control in a car radio or stereo system.

Alternatively, an additional single window in an optical wheel can activate a *home sensor*. When the device is powered up, the wheel is turned until the home sensor is triggered, at which point the orientation of the wheel is known, and subsequent pulses from rotation sensor(s) will add or subtract angular information.

The pulse generated by a home sensor may be described in datasheets as a *reference signal* or *index signal*. In the early days of desktop computers, each 5.25-inch floppy diskette was perforated with an index hole for this purpose.

The Gray Code

For greater reliability in determining absolute position, an optical wheel can be divided into several concentric tracks, each of which contains a different coded sequence and is assigned its own light emitter and light sensor. The sensors are arranged in a radial line to scan the disc as it rotates. Since each detector will provide either a signal or no signal, output from the set of sensors can be combined to create a binary number.

Figure 7-10 shows binary codes from 0000 through 1111, corresponding with decimal numbers 0 through 15, where a white square is equivalent to a numeral 1 and a black square is equivalent to a numeral 0.

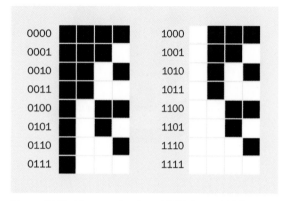

Figure 7-10 *Binary codes from 0000 through 1111, using white to represent 1 and black to represent 0.*

Figure 7-11 shows this system mapped onto a codewheel. The red circles indicate the locations of four stationary light detectors, which begin by providing a reading of 0000, since they coincide with four opaque areas of the wheel. Now if the wheel makes 1/16th of a full rotation in the direction of the arrow, the detectors will register 0001. If the wheel continues to rotate, the detectors will count in binary up to 1111 before the sequence repeats.

The problem with this design is that small manufacturing inaccuracies and other imperfections will result in some light detectors responding fractionally more quickly than others as the wheel rotates. This will occur in transitions where two or more adjacent segments of the wheel change between transparency and opacity. For instance, where 0011 is followed by 0100, transient values of 0010, 0001, 0111, 0110, or 0101 are possible. Although they will be brief, these values may trigger associated electronics.

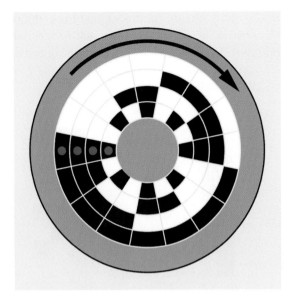

Figure 7-11 *The binary sequence mapped onto a code-wheel as areas of opacity and transparency. Red circles indicate light detectors.*

To eliminate this problem, a different code sequence can be used in which only one of the four sensors is allowed to make a transition from each value to the next. This is called a *Gray code*, and eliminates the issue of simultaneous transitions. A commonly used Gray code is shown in Figure 7-12.

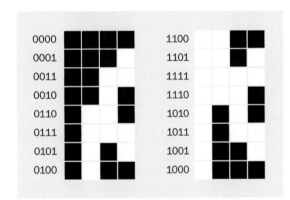

Figure 7-12 *A Gray code that allows only one binary digit to change in each transition from one value to the next.*

Magnetic Rotary Encoders

If a ferrous wheel is magnetically polarized in multiple domains, its rotation can be detected

by a Hall-effect sensor in much the same way that a wheel divided into transparent and opaque sections can be assessed by light detectors and emitters. This is illustrated in Figure 7-13, where magenta and cyan bands indicate north and south magnetic poles.

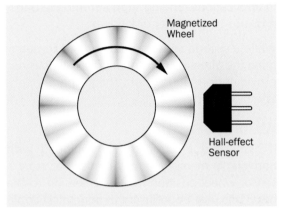

Figure 7-13 *A Hall-effect sensor can detect the rotation of a wheel that is divided into multiple north and south magnetic poles.*

See "Hall-Effect Sensor" for information about Hall-effect sensors.

An additional Hall sensor can be added, offset from the first, as in Figure 7-14. Once again the phase difference between the pulse trains can be used to determine the direction of rotation.

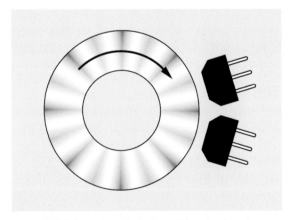

Figure 7-14 *Direction of rotation of the wheel can be deduced from the phase difference between the pulse trains from two sensors.*

Alternatively, a toothed wheel can be used, as suggested in Figure 7-15.

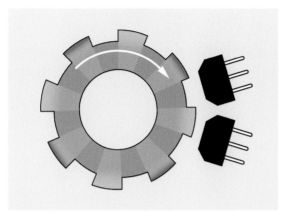

Figure 7-15 *A toothed wheel can trigger a Hall sensor if the teeth are magnetized.*

Another option is to mount a pair of magnets on a nonmagnetic wheel, with a Hall-effect sensor in the center. If it is a sensor with an analog output, the voltage will fluctuate smoothly between positive and negative, relative to the power supply for the sensor. Alternatively, a bipolar Hall-effect sensor can be used to provide a binary output. The concept is illustrated in Figure 7-16. An advantage of this configuration is that it provides approximate information about the absolute position of the wheel.

The two magnets will provide usable linear sensor outputs over a span of plus-or-minus 30 degrees of rotation, approximately. Additional magnets or sensors can produce a more complex output that would be decoded by a microcontroller.

How to Use It

An optical or rotary encoder is well suited for use with a microcontroller program that can count pulses, compare pulse trains, or interpret a Gray code. The microcontroller then takes appropriate action. For example, if a rotational encoder is used to control the gain of an audio amplifier, the microcontroller determines the direction and angle through which the encoder

turns, and can respond by changing the value of a **digital potentiometer** (see Volume 1).

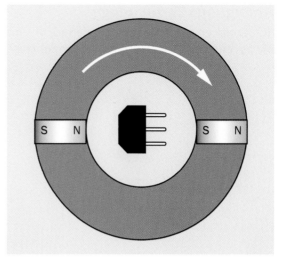

Figure 7-16 *A Hall-effect sensor can detect the angle of rotation of a ring on which two magnets are mounted as shown.*

Integrated circuit chips are available to convert a sequence of quadrature signals into an up-pulse or a down-pulse, thus eliminating the chore of achieving this with a microcontroller program. The LS7183 by LSI Computer Systems is an example.

What Can Go Wrong

Wiring Errors

If two sensors are used to detect the direction of rotation of a rotary encoder, and outputs from the sensors are accidentally swapped, the component will seem to work normally except that the indicated direction will be inverted.

Coding Errors

Microcontroller code that interprets quadrature signals must be fast enough to keep up with the pulse streams. If a microcontroller is performing other tasks, an interrupt may be necessary for processing the rotational data. This should not be a significant problem when interpreting human input from a knob or dial, but for a motor-driven encoder, pulse counting in hardware may be a better alternative.

Ambiguous Terminology

Rotational encoders are often referred to simply as "encoders," because they are more common than linear encoders. When searching for an encoder, check each datasheet carefully to determine which type you are dealing with.

tilt sensor

A *tilt switch* is defined here as an electromechanical switch, while a *tilt sensor* uses electronics. Both types are included in this entry.

A *tipover switch* is very similar to a tilt switch and uses the same principles. Therefore, it is included in this entry.

Some manufacturers' datasheets refer to tilt sensors as *tip sensors*. Some Asian supply catalogues refer to tilt switches as *breakover switches*.

An **accelerometer** can measure the angle at which it is held relative to the downward force of gravity, but it has additional capabilities. Therefore, it has its own separate entry in this Encyclopedia.

An *inclinometer* measures the incline, or positive slope, from an observation point to the top of an object such as a building or tree. The height of the object can be calculated from the angle. A *clinometer* can additionally measure a decline, or negative slope. These measurement devices are fully featured products as opposed to sensors, and are not included in this Encyclopedia.

OTHER RELATED COMPONENTS

- **accelerometer** (see Chapter 10)
- **vibration** sensor (see Chapter 11)

What It Does

Three principal types of tilt sensor exist.

1. Single axis, single output. The sensor responds to being tilted around one horizontal axis, relative to the downward force of gravity.

2. Dual axis, dual output. The sensor contains two sensing elements at 90 degrees to each other. Each has an output determined by its angle of tilt from vertical around one axis.

3. Dual axis, single output. A single sensor responds to an angle of tilt from vertical around any horizontal axis.

A *tilt switch* is usually of the third type, and is defined here as containing an electromechanical or electronic switch that opens or closes a connection. Most tilt switches are SPST (normally open) or SPST (normally closed). A minority are DPDT.

A *tipover switch* is a type of high-current tilt switch that cuts power to a device such as an electric heater when it is tipped over.

This Enclopedia defines a *tilt sensor* as being an electronic component, as opposed to an electromechanical component. The distinction is often observed in datasheets, but not always.

Schematic Symbol

No specific schematic symbol is generally used for any variant of a tilt sensor. It can be represented with an annotated switch symbol.

How It Works

Because a tilt switch is a simpler device than a tilt sensor, it will be described first.

The most common type of tilt switch consists of a cylindrical metal or plastic enclosure, often measuring about 5mm by 15mm, containing two spherical steel balls that may be nickel-plated or gold-plated. When the switch is tilted, the balls eventually run downhill, and the lower ball completes an electrical connection between two contacts or between a single contact and the metal enclosure of the switch. The second ball is included to add weight and suppress vibration in the first.

Figure 8-1 shows a switch manufactured by Comus Global, rated for 0.25A at 60VAC or 60VDC, maximum. The body of the switch measures approximately 10mm x 5mm, and the switch is activated when tilting -10 degrees from horizontal. It is deactivated when tilting +10 degrees. A scale drawing of the interior is shown in Figure 8-2.

Figure 8-1 *The CW1300 tilt switch manufactured by Comus Global.*

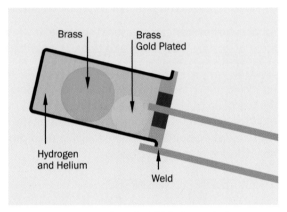

Figure 8-2 *Interior of the CW1300 tilt switch, from a scale drawing supplied by the manufacturer. The lower lead is welded to the shell of the sensor. The leads may be inserted in a circuit board.*

Figure 8-3 shows three common internal configurations of a generic tilt switch. The top version has axial leads and uses the metal shell of the switch to complete the circuit. The center version has radial leads, with a plastic shell. The bottom version has radial leads, one of which is attached to the metal shell to complete the circuit.

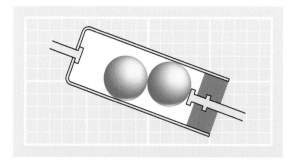

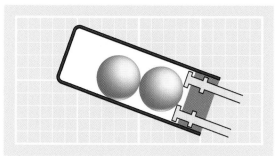

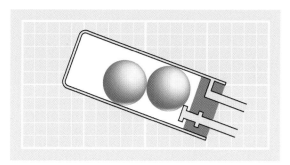

Figure 8-3 *Three variants of a generic ball-operated tilt switch. The graph-paper scale is in millimeters.*

Figure 8-4 shows the parts of a disassembled tilt switch.

Simplified Version

Where size is not a significant consideration, a tilt switch can be created by attaching a pivoting, weighted arm to a small snap-action switch.

Applications

An old-style (nonelectronic) thermostat may contain a tilt switch attached to the end of a bimetallic strip coiled into a spiral. When the strip bends in response to a drop in temperature, the switch closes its contacts, activating a relay that starts a heating unit. If the temperature rises, an additional set of contacts in the same relay may activate an air-conditioning unit. In old thermostats, the tilt switch may contain mercury in a glass tube, which should be handled with caution.

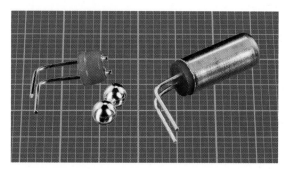

Figure 8-4 *At right, a tilt switch. At left, the cap removed, and the two balls that make internal contact. The background grid is in millimeters.*

A tilt switch may detect the opening of a door or window in a simple alarm system.

Tilt switches have been used in automobiles to switch on the interior light in the trunk when its lid is opened.

A normally closed tilt switch is often used to stop the inflow of granular material to a bin when it is almost full. This is colloquially known as a *bin switch*. In industrial applications of this kind, the switch is activated by a long lever that has a ball mounted at the end. The switch assembly is physically large. See Figure 8-5.

A normally open tilt switch may operate a valve or start a pump when the liquid in a tank drops below a certain point. If the switch uses a float to sense the level of the liquid, it is often known as a *float switch*. This is described in the entry on liquid level sensors. See Chapter 15.

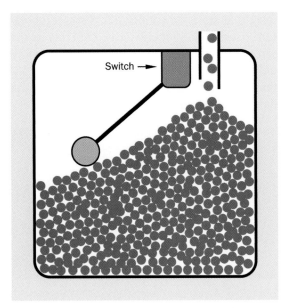

Figure 8-5 *The flow of granular material into a bin can be sensed and stopped by a tilt switch of this type, often known as a bin switch.*

A *tipover switch* may use the simplified system of a weighted arm that activates a snap-action switch. When used in conjunction with a room heater, the switch must handle substantial current.

A motorcycle may contain a tipover switch to stop the electric fuel pump if the motorcycle falls on its side.

Four tilt switches placed in a cross-shaped pattern on a flexible mount can be used as the basis of a very simple game controller, with a joystick mounted in the center.

Variants

The three configurations of ball-type tilt switches shown in Figure 8-3 are functionally interchangeable and can be chosen for convenience of their leads in fitting the circuit.

Mercury Switches

Early tilt switches contained a blob of mercury in a glass bulb. When the bulb was tilted, the mercury rolled to the end and made an electric

connection between two metal contacts that penetrated the bulb.

A small mercury switch is shown in Figure 8-6. This type of sensor became less common after many countries classified mercury as an environmental hazard and established regulations restricting its use.

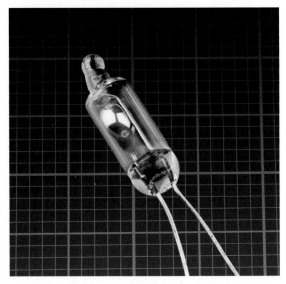

Figure 8-6 *Small mercury switch rated for 0.3A at 24VDC or 24VAC. Larger mercury switches can switch more power; 1A at 230V is common. The background grid is in millimeters.*

Mercury is an excellent electrical conductor. It remains in a liquid state between about -38 degrees Celsius and +356 degrees Celsius, and has very high surface tension, encouraging it to form a single blob instead of breaking up into small droplets. Because free space in the bulb is filled with an inert gas to prevent oxidation of the electrodes, a mercury switch can have a very long operating life. In the United States in the 1970s, some light switches were sold containing mercury switches with a claimed life expectancy of 100 years.

Pendulum Switch

This type of switch, now relatively rare, was found in vintage pinball machines. It consists of a pendulum about 5cm long, suspended inside a steel ring about 1cm in internal diameter. If the machine was rocked sufficiently to bring the pendulum in contact with the ring, the game was cancelled and the word "Tilt" appeared on the display. Consequently this was referred to as a tilt switch, although really it was a form of **vibration** sensor with a long period of oscillation.

Magnetization

Some tilt switches use a steel ball that is weakly magnetized, so that it will seat itself more firmly when it rolls into a circular depression or ring. This type of switch must be tilted back through a larger angle to dislodge the ball. Therefore it will exhibit greater *hysteresis*.

Tilt Sensors

Unlike a tilt switch, a tilt sensor is not built around an electromechanical switch.

The principle of a rolling ball has been miniaturized and encapsulated in a small enclosure (10mm square or smaller), in which the ball rolls to interrupt a beam from an internal LED shining on a phototransistor. Examples are found in the Panasonic AHF series. Internal circuitry ensures a clean on-off signal, free from the switch bounce that is a problem in basic ball-type tilt switches. However, the switch requires a power supply, and the open-collector output must be used with a pullup resistor. By comparison, a simple electromechanical tilt switch can be wired directly to the device that it controls.

Diagrams from the Panasonic datasheet show the three types of AHF sensor available for vertical, horizontal, and reverse mounting. In each case, the ball (dotted circle) rests in a shallow cup (dotted curve) where it obstructs the beam from an internal LED (not shown). See Figure 8-7. An exterior view of the AHF22 is shown in Figure 8-8.

Vertical mounting	Horizontal mounting	Reverse mounting
AHF21	AHF22	AHF23

Figure 8-7 *Three types of Panasonic tilt sensor, from the manufacturer's datasheet. See text for details.*

Figure 8-8 *Exterior view of the Panasonic AHF22 tilt sensor. The background grid is in millimeters.*

Two-Axis Tilt Sensors

The Rohm RPI-1035 is a surface-mount tilt sensor about 4mm square, with two phototransistor outputs that indicate which axis the sensor is tilting around. The outputs can be interpreted as a 2-bit binary number, with its four possible states indicating the rotation of the switch around two axes at 90 degrees to each other. Switches of this type were developed to indicate the orientation of consumer-electronic devices such as digital cameras, but more sophisticated sensors containing accelerometers are becoming price-competitive.

Surface-mount 2-axis tilt sensors have been made available on small breakout boards that are easy to use experimentally. An example is the Parallax 28036 shown in Figure 8-9.

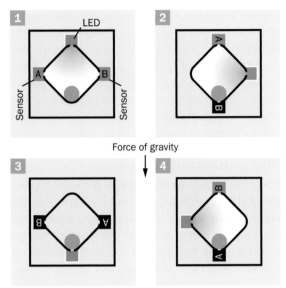

Figure 8-10 *The rolling-ball sensor inside the Parallax 28036. See text for details.*

Figure 8-9 *A 2-axis tilt sensor mounted on a breakout board available from Parallax.*

The behavior of the rolling-ball sensor at the heart of this board is shown in Figure 8-10. The sensor contains a square cavity in which the ball is depicted as a blue circle. At one corner of the cavity is a red LED, while two phototransistors, labeled A and B, are at the corners on the left and right. When the sensor is held as in section 1 of the figure, with the LED at the top and the ball resting at the bottom, both phototransistors have a high output as they receive light from the LED.

In section 2 of Figure 8-10, the sensor has been turned through 90 degrees. The ball now prevents light from reaching sensor B, while phototransistor A is still active. In section 3 of the figure, the sensor has been turned through another 90 degrees, so that the ball prevents light from escaping from the LED, and both phototransistors are now dark. In section 4, the ball obstructs phototransistor A but not phototransistor B.

Suppose that the sensor is placed flat on a horizontal surface. The sensor will now respond to being tipped either way around two horizontal axes. This justifies its description as a 4-directional tilt sensor, although its design suggests that it may have been intended for use as described above, rotating through four positions around one horizontal axis.

Values

A heavy-duty tilt switch can be rated as highly as 10A at 240VAC. More commonly, a tilt switch about 15mm long can be expected to switch about 0.3A at 24VAC or 24VDC.

The *operating angle* is the angle through which the switch must be turned to activate it, relative to its normal rest position.

The *return angle* is the angle to which the switch must be returned to deactivate it. Hysteresis results from the return angle being smaller than the angle that activates the sensor.

Tilt sensors with an open-collector output will specify maximum forward current for the internally mounted LED (usually no greater than

50mA), the maximum collector-emitter voltage at the output (typically 30V), and the maximum collector current (often 30mA). An explanation of open-collector outputs is given in Appendix A (see "3. Analog: Open Collector").

How to Use It

An electromechanical tilt switch can be connected directly between a power supply and a device, so long as the device does not draw more current than the switch is rated to handle. Note that inductive loads such as motors draw an initial surge that can be at least twice the operational rating, while relays may be more likely to create a voltage spike when disconnecting. Switches should be chosen accordingly. For more discussion of this topic, see **switch** in Volume 1.

A small tilt switch can be used in conjunction with a relay or transistor to amplify its signal sufficiently to drive a larger load.

If an electromechanical tilt switch is connected to an electronic device such as a microcontroller or logic chip, output from the switch will have to be *debounced* to prevent a series of brief voltage spikes that can cause false triggering when the switch turns on or off. A debouncing logic circuit or chip can be used, or the program code in a microcontroller can introduce a wait period of up to 50 milliseconds to allow the contacts to settle.

Mercury switches are much less likely to create a noisy output than rolling-ball switches, and may require little or no debouncing.

For an application that must sense rotation around two or three axes, multiple single-axis tilt switches can be combined. A microcontroller or logic gates will be necessary to evaluate signals from the switches, to determine the orientation.

What Can Go Wrong

Contact Erosion

If a ball-type tilt switch is subjected to current that exceeds its specification, arcing may erode its contacts, and they will become less reliable, especially if the contacts are plated with a thin metallic film that is eroded. For additional information on arcing in switches, see the **switch** entry in Volume 1.

Random Signals

During the brief time when a ball-type tilt switch is turning from one position to the other, vibration of the ball(s) inside it is likely to create erratic, random signals. If the output from the switch is being evaluated by a microcontroller, a debouncing routine may be insufficient to prevent the random signals from being sensed, and some programming will be necessary to ignore the signals during this transitional phase. If the switch is connected directly to a relay, the intermittent signals may occur sufficiently rapidly that the relay will ignore them.

Environmental Hazard

A device that incorporates a mercury switch may have to be re-engineered in the future if the availability of mercury switches becomes unreliable as a result of tighter environmental regulations. For the same reason, the end user may have difficulty replacing a mercury switch if it fails. Therefore, a ball-type tilt switch should be used instead of a mercury switch in any newly designed device.

Requirement for Gravity

Because a tilt switch depends on gravity to roll a ball or move a blob of mercury, it will not work in low-gravity, reversed-gravity, or zero-gravity conditions—for example, in a rocket during the unpowered phase of ascent and descent, or in an aircraft that performs aerobatic maneuvers. Performance of a tilt switch in a vehicle that accelerates or decelerates suddenly may also be unreliable. Likewise, it cannot be used on a small boat.

Requirement for Stability

A tilt switch will tend to give erroneous results in a location where there is significant vibration or where the object containing the switch may be turned or repositioned unpredictably by the user.

gyroscope

Historically, a **gyroscope** always contained a spinning disc. Some devices for navigation still depend on rotating elements, but they are outside the scope of this Encyclopedia. This entry deals primarily with *vibrating gyroscopes*, also known as *resonator gyroscopes*, that are MEMS devices contained within silicon chips.

OTHER RELATED COMPONENTS

- **accelerometer** (see Chapter 10)
- **GPS** (see Chapter 1)
- **magnetometer** (see Chapter 2)

What It Does

A **gyroscope** resists rotation around any axis at right angles to its own axis of rotation or vibration. Consequently, if the gyroscope is allowed to move freely on gimbals in a sealed enclosure, the gyroscope will tend to maintain its orientation while the enclosure can rotate freely around it.

Taking this concept a step further, if the enclosure is mounted in an aircraft, the aircraft's rotation around two axes can be determined by referring to the gyroscope. If additional gyroscopes are added orthogonally to the first, the aircraft's rotation around all three axes can be determined.

A gyroscope does not measure linear motion in any direction, or any static angle of orientation.

Schematic Symbol

A chip-based gyroscope, magnetomer, or accelerometer may be represented in a schematic as a rectangular box containing abbreviations to identify pin functions (as in any integrated circuit chip).

IMU

An **accelerometer** measures variations in linear motion and will also measure its own static orientation relative to the force of gravity. If an accelerometer rotates around its own axis, it will not measure angular velocity.

A **magnetometer** measures the magnetic field surrounding it, and may be sufficiently sensitive to determine its orientation relative to the Earth's magnetic field.

When an accelerometer and a gyroscope are contained in the same package, optionally with a magnetometer, they may be described as an *IMU* (inertial measurement unit), which can provide necessary data to maneuver aircraft, spacecraft, and watercraft, especially when **GPS** signals are unavailable.

Applications

The first chip-based gyroscope was used in automobiles in 1998 as a yaw sensor in a skid-control system. Subsequent automotive applications include active suspension control, air bag sensors, rollover detection and prevention, and navigation systems.

Gyroscopes may be installed in military ordnance to provide backup in case an onboard **GPS** system fails, possibly as a result of radio jamming.

Handheld 3D game controllers and headsets may use gyroscopes to control images displayed to the viewer. A digital camera may employ a gyroscope to provide image stabilization. Gyroscopes are usually found in quadcopters or drones, are used to stabilize two-wheeled vehicles such as the Segway, and are used in robotics.

How It Works

The traditional form of gyroscope is a rotating wheel, which will resist turning forces perpendicular to its own axis of rotation. In Figure 9-1, three directions at right angles to each other are defined in the bottom-right corner of the diagram as X, Y, and Z. The wheel is rotating around the X axis, as shown by the green arrow. It will resist any turning force around the Y axis (red arrows) or the Z axis (yellow arrows).

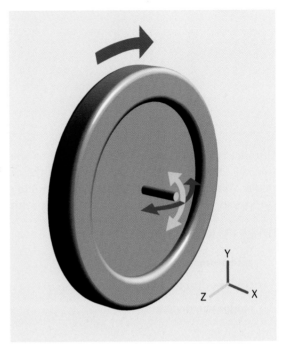

Figure 9-1 *In a traditional gyroscope, a wheel that is rotating (shown by the green arrow) will resist a turning force on either of the axes (shown by red and yellow arrows) perpendicular to its axis of rotation.*

Vibrating Gyroscope

A vibrating fork can be substituted for a wheel. In Figure 9-2 a fork is secured at its base while its tines are induced to vibrate toward each other and away from each other, as suggested by the double-ended arrow. In a chip-based gyroscope, this vibration is induced piezoelectrically or by static electricity.

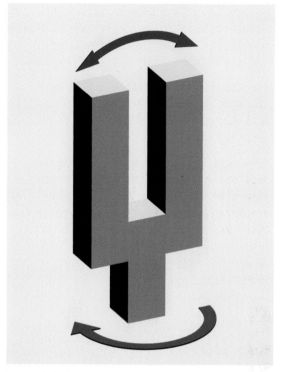

Figure 9-2 *A vibrating fork (green arrows) can be substituted for a rotating wheel in a gyroscope.*

Figure 9-3 *A turning force is applied to the base of the fork around a vertical axis, as shown by the lower arrow.*

Now suppose a turning force is applied around the vertical axis at the base of the fork, as suggested by the lower arrow in Figure 9-3.

The angular momentum of the vibrating tines causes them to resist this turning force, and consequently they will tend to bend, as shown by the yellow arrows in Figure 9-4. The amount of their deflection can be measured capacitively. This system is used in many chip-based gyroscope systems, and may be referred to as a *vibrating gyroscope* or a *resonator gyroscope*.

Figure 9-4 *The angular velocity of the rotating fork causes deflection of its vibrating tines, shown by the yellow arrows.*

An assembly of microscopic forks can be etched into a silicon chip. In Figure 9-5, an electron micrograph shows the interior of this type of chip. It contains three rotational sensors, responding to the X, Y, and Z axes of motion. These sensors respond to *pitch* (rotation around the X axis), *roll* (rotation around the Y axis), and *yaw* (rotation around the Z axis).

Fork-based sensors are analog devices whose values are converted to digital values by an onboard *analog-to-digital converter* (ADC). The values are stored in registers that are available to other devices, often via the I2C protocol, which is widely used by microcontrollers.

For additional details about I2C, see Appendix A.

Typically there will be two 8-bit registers for each axis. Each register stores the binary equivalent of a signed integer, where the positive or

negative value represents the direction and magnitude of deflection, usually in degrees per second (dps).

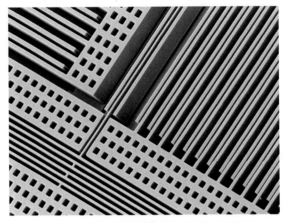

Figure 9-5 *Electron micrograph of the STMicroelectronics LIS331DLH vibrating gyroscope installed in the Apple iPhone 4. The parallel plates at bottom-left function as a spring, while the elements at top-left and at right measure capacitance as their orientation varies according to rotational velocity. Photo courtesy of MEMS Journal published by Chipworks (http://www.chipworks.com)*

Variants

The L3G420D by STMicroelectronics is a 3-axis gyroscope-only chip. It communicates via the SPI or I2C protocol, is approximately 4mm square, and can measure rotational rates up to plus-or-minus 2,000 degrees per second.

The Freescale FXAS21002C has a similar specification. Gyroscope-only chips of this type have fallen in price to the point where they are comparable to the retail cost of everyday components such as a small-signal relay or an audio amplifier on a chip.

IMUs

Chips that only contain gyroscopes are becoming less common as the cost of adding accelerometers decreases.

Gyroscopes and accelerometers are complementary, as gyroscopes are insensitive to linear motion or the Earth's gravity, but accelerometers can measure the rate of change of linear motion and orientation of the chip relative to the Earth. Software can combine this data to calculate the shape of the path being described by a device containing the chip, in addition to the changing velocity of the chip along that path.

The InvenSense MPU-6050 is a common 3-gyroscope, 3-accelerometer chip. It also includes an interface for connecting an external 3-axis magnetometer. The SPI and I2C communications protocols are supported. The MPU-6050 has been a popular choice in the hobby-electronics community, so that Arduino-compatible code to interpret its data is available from many sources. Breakout boards are available with the MPU-6050 installed. An example is the Sparkfun SEN-11028 shown in Figure 9-6.

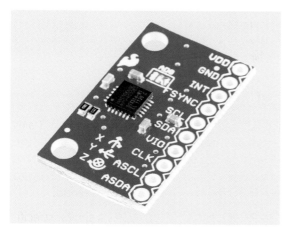

Figure 9-6 *A breakout board from Sparkfun, providing easy access to the InvenSense MPU-6050 chip combining three gyroscopes and three accelerometers.*

Values

The *rotational velocity* of a gyroscope element is usually expressed in degrees of rotation per second (dps), and sometimes in rotations per minute (RPM).

A datasheet will specify the number of sensor axes (usually 3), supply voltage (3.3VDC is common), maximum digital-low and minimum digital-high output voltages, and power consumption in normal mode and sleep mode. Power consumption is usually less than 10mA.

The *dynamic range* is the maximum forward and reverse rotational velocity, which usually will not exceed plus-or-minus 2,000 degrees per second. Lower ranges may be user-selectable. The advantage of selecting a lower maximum rate of change is that it can be converted to a digital value with higher precision.

The *sensor resonant frequency* will be several kilohertz, and must be higher than the frequency of any vibration that is applied to the sensor during use.

The *communications protocol* is usually I2C and SPI may be offered as an option, with a selectable digital output data rate.

Bias temperature coefficient describes the effect of temperature on the gyroscope.

The *resolution* of the gyroscope relates to the number of bits used in the digital output from the onboard ADC. A 16-bit resolution is common.

How to Use It

Using a smart chip such as the MPU-6050, the circuit designer can take advantage of its onboard digital motion processor (DMP). Still, obtaining orientation information from the contents of registers on the MPU-6050 is non-trivial. Online sources and code libraries are necessary. The book *Make: Sensors* contains code listings for the Raspberry Pi as well as the Arduino.

What Can Go Wrong

Temperature Drift

Vibrating materials at the heart of a chip-based gyroscope are likely to change their behavior with temperature. Typically the chip will contain a temperature sensor, the value of which can be used to adjust the output value of the gyroscope.

Mechanical Stress

Stress can be induced when a surface-mount chip is soldered to a board. The vibrating parts of a chip-based gyroscope may be adversely affected. Datasheets will supply information regarding maximum acceptable temperature during the soldering process.

Vibration

Because a chip-based gyroscope depends on the consistent behavior of internal vibrating parts, external vibration can degrade its accuracy. Sensor design can minimize the effects of vibration, but the datasheet should be consulted for details.

Placement

A gyroscope should be placed on a circuit board near a hard mounting point where deflection of the board will be minimized.

accelerometer

10

OTHER RELATED COMPONENTS

- **GPS** (see Chapter 1)
- **gyroscope** (see Chapter 9)
- **tilt** sensor (see Chapter 8)
- **vibration** sensor (see Chapter 11)

What It Does

Acceleration is the rate at which velocity changes over time. If a car takes 10 seconds to increase its speed from 30kph to 40kph relative to the road on which it is traveling, it is accelerating at an average of 1kph each second. If it then reduces its speed back to 30kph during another 10-second interval, it is decelerating at the same rate—although deceleration is really just negative acceleration.

While a car is accelerating, passengers will feel a lateral force exerted on them. Similarly, astronauts in a rocket that blasts off will feel a downward force. According to Einstein's theory of equivalence, forces resulting from acceleration are indistinguishable from the force of gravity.

Consequently, a sensor that measures acceleration can also measure gravity. This sensor is an **accelerometer**. Its output may be measured in *gravities*, abbreviated with the letter g (not to be confused with the usage of G to measure the strength of a magnetic field in gauss).

If three accelerometers are mounted orthogonally (at 90 degrees to each other), their readings can reveal:

- The direction of acceleration of a moving object.
- If an object has been dropped and is falling freely.
- Which way up it is being held in a stationary position.
- The severity of an impact when a moving object collides with some other object.

IMU

A **gyroscope** measures the rate of rotation of the enclosure in which it is mounted. This is properly known as the *angular velocity*. A gyroscope will also respond to changes in the rate of rotation. It does not measure linear motion or a static angle of orientation.

A **magnetometer** measures the magnetic field surrounding it, and may be sufficiently sensitive to determine its orientation relative to the Earth's magnetic field.

When an accelerometer and a gyroscope are contained in the same package, optionally with a magnetometer, they may be described as an *IMU* (inertial measurement unit), which can provide necessary data to maneuver aircraft, spacecraft, and watercraft, especially when **GPS** signals are unavailable.

Schematic Symbols

A chip-based accelerometer may be represented in a schematic as a rectangular box containing abbreviations to identify pin functions, as in any integrated circuit chip. No specific symbol is used for any of these components.

Applications

In the past, accelerometers were laboratory devices that calibrated the performance of cars, airplanes, and other types of vehicles. Measuring the ability of car tires to withstand cornering forces was an application where an accelerometer was used.

Miniaturization of accelerometer elements, coupled with a radical reduction in their cost, has enabled them to be installed in small electronic devices ranging from smartphones to hard drives.

In a phone or a camera, accelerometers can determine which way up the user is holding the device. The camera can rotate the picture appropriately, and the orientation of the picture can be saved with its image data.

In an external hard drive containing rotating platters, accelerometers can protect the read-write heads by rapidly parking them during the fraction of the second that elapses between someone dropping the hard drive and its impact with the floor.

Accelerometers can be installed in a 3D mouse or virtual-reality headset to determine its orientation and motion. This enables a video image to be updated appropriately. For example, the Nintendo Wii Remote has been marketed with an ADXL330 accelerometer.

In an automobile, an accelerometer can trigger the deployment of an air bag when the deceleration caused by an impact exceeds a threshold level.

How It Works

The simplest conceptual model of an accelerometer consists of a mass attached to one end of a coiled compression spring. The other end of the spring is anchored in an object whose acceleration is being measured. The mass can only move along the same axis as the spring.

Figure 10-1 shows three views of a simplified accelerometer, which is a sealed tube shown in dark red. In the center image, the accelerometer is in its rest state. The top image shows the mass (a dark blue square) responding when the tube accelerates from left to right. The third image shows it decelerating (that is, undergoing negative acceleration, or acceleration from right to left). Using an ideal spring, the displacement of the mass will be proportional to the rate of acceleration, within reasonable limits. The displacement can be measured optically or capacitively.

Note that the rest state will resume when the acceleration stops, regardless of constant motion in any direction. An accelerometer only measures a *change* in velocity. It does not measure a constant velocity.

An accelerometer cannot measure rotation around its own axis of movement. Therefore, it may be used in conjunction with a **gyroscope**, which measures angular velocity.

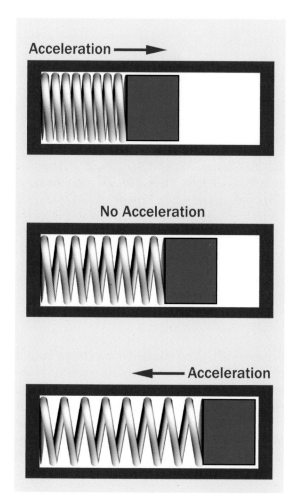

Figure 10-2 *Left: an accelerometer resting on the ground measures 1g, as the force of gravity pulls the mass downward. Right: in free fall, the accelerometer measures 0g.*

If the device is dropped so that it falls freely under the influence of gravity, it is in *free fall* and will accelerate at approximately 9.8 meters per second each second. This is usually written as 9.8m/sec^2, or can be described as 1 gravity, often expressed as 1g.

An accelerometer in free fall will measure 0g, as shown on the right in Figure 10-2, because all parts of the accelerometer are now accelerating equally under the force of gravity.

If three accelerometers are assembled orthogonally, and if one accelerometer is vertical, and the device is held motionless relative to the Earth, the vertical sensor will show 1g while the other two accelerometers will show 0g. If the device is dropped, all sensors will show 0g.

Rotation

If a device containing three accelerometers is turned over, and is not in free fall, accelerometers mounted orthogonally will show values that vary depending on their alignment with the force of gravity.

Figure 10-1 *Simplified view of an accelerometer consisting of a sealed tube (dark red outline) with a spring anchored to it at left. A small mass (dark blue) is attached to the right end of the spring. The mass responds to acceleration of the tube.*

Gravity and Free Fall

If a device containing the simplified accelerometer described previously rests on the ground, and is mounted vertically as in the left section of Figure 10-2, gravity acting on the mass will apply a force to one end of the spring, while the other end is restrained. The accelerometer will now measure 1g as a downward force.

Calculation

The force acting upon an object will cause a rate of acceleration that can be calculated by Newton's Second Law of Motion, provided that the object can move freely and is not being subjected to any additional forces, such as the force of gravity. If F is the force, m is the mass of the object, and a is the acceleration:

    F = m * a

And therefore:

    a = F / m

If the mass is restrained by a spring whose compression or extension has an approximately linear relationship to the force applied to it, acceleration can be calculated as a function of the linear displacement of the mass.

These statements ignore relativistic effects that are insignificant unless ultra precise time and motion measurement may be involved.

In a real-world accelerometer, movement of the mass will require some form of damping to prevent it from oscillating.

Variants

Accelerometer prices dropped radically after 2010. In an effort to maintain profitability, manufacturers have loaded more features onto chips. While a 2-axis accelerometer such as the Memsic 2125 seemed a good option when first introduced, it is now facing obsolescence as 3-axis accelerometer chips that also contain 3-axis gyroscopes have become ubiquitous—and no more expensive.

Early chip-based accelerometers provided analog outputs where voltage was proportional with acceleration and could be processed by a comparator. On some breakout boards, such as the Dimension Engineering DE-ACCM6G, which used the STMicroelectronics LIS244ALH chip, a comparator was included (see Figure 10-3).

Figure 10-3 *A relatively early, relatively expensive breakout board using the 2-axis accelerometer LIS244ALH chip by STMicroelectronics. It has been superceded by chips that combine accelerometers, gyroscopes, and processors to provide a digital output.*

Because this board allowed a maximum output of only 0.83mA, it was only suitable for high-impedance logic chips or a microcontroller. However, because the output was analog, it could be passed through another comparator for direct connection to a piezo beeper to create a device that would sound an alarm when tilted. This is shown in Figure 10-4.

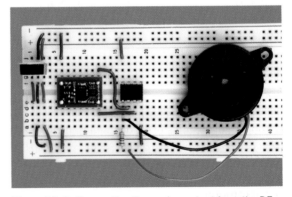

Figure 10-4 *Connecting the analog output from the DE-ACCM6G 2-axis comparator breakout board through a comparator to a piezo beeper.*

Many chips now contain gyroscopes as well as accelerometers. The electron micrograph in Figure 10-5 shows the interior of a chip of this kind, where the zig-zag shapes are "springs" etched into the die, the large areas patterned with dots are masses that can respond to various forms of motion, and the parallel plates are capacitive sensors.

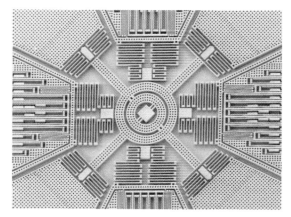

Figure 10-5 *Interior of a chip combining gyroscopes with accelerometers.*

The complexity of a chip combining two types of sensors creates a need for more complicated code to process the six outputs. Consequently most accelerometer chips now have their own onboard ADCs, and digital registers for communication with microcontrollers via the I2C communication protocol. Some chips also have onboard processing to interpret the mix of data. However, microcontrollers still need code to make sense of the data.

The hobby-electronics community has responded. A breakout board such as the LSM9DS0 from Adafruit, using a chip from STMicroelectronics that shares that same part number, tries to make this extremely complex chip accessible to experimenters. Shown in Figure 10-6, the LSM9DS0 contains a 3-axis magnetometer in addition to a 3-axis gyroscope and a 3-axis accelerometer.

The datasheet for the LSM9DS0 runs to more than 70 pages. At the time of writing, Adafruit is still refining code for the Arduino to make the features of this chip accessible.

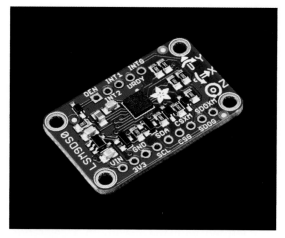

Figure 10-6 *This breakout board from Adafruit is built around an LSM9DS0 chip that combines accelerometers with gyroscopes and magnetometers.*

Despite the complexity of the LSM9DS0 chip, the breakout board sells for approximately the same price, at the time of writing, as the DE-ACCM6G 2-axis analog-output magnetometer four years ago. In the future we may expect IMU prices to fall still further, so that this type of multifunction chip becomes the default, and users simply ignore the functions that they don't need.

Values

Current consumption will vary depending on the activity of the chip, and may be broken down for different types of sensors on multi-sensor chips. A modern accelerometer may typically draw less than 1mA. Gyroscope power consumption will be greater, as a segment of the chip will be maintained in a state of vibration.

Linear acceleration measurable by an accelerometer is customarily expressed in gravities, abbreviated g. Current generations of chips may be able to measure as much as plus-or-minus 16g, but because the value is converted internally to a digital quantity, smaller accelerations will not be expressed so accurately. Therefore the measurement range for acceleration is usually user-selectable by sending an appropriate code to the chip. Ranges may include plus-or-minus 2g, 4g, 6g, 8g, and 16g.

Sensitivity defines the smallest increment of acceleration measurable by the least significant bit (LSB) in an output register, for each of the acceleration ranges. In a range of plus-or-minus 2g, the internal 16-bit ADC may be capable of measuring 0.06 milligravities. When the range is reset to plus-or-minus 16g, the smallest increment may be around 0.7 milligravities.

The measurable range of gravities should not be confused with the maximum acceleration that the chip can tolerate without suffering internal mechanical damage, either while it is powered or unpowered. This will be more than 1,000g provided the duration is brief, and would only be experienced during an impact.

Linear acceleration sensitivity change versus temperature is usually expressed as a percentage, such as plus-or-minus 1.5%.

Output type will be analog or digital. If digital, the data protocol will be I2C or SPI. If I2C, the address of the device should be configurable. The data transfer rate is typically at least 100kHz, and this also may be configurable to different rates.

For additional details about protocols such as I2C, see Appendix A.

What Can Go Wrong

Mechanical Stress

Stress can be induced when a surface-mount chip is soldered to a board. The moving parts in a chip-based accelerometer may be adversely affected. Datasheets will supply information regarding maximum acceptable temperature during the soldering process.

Other Problems

If the accelerometer function is combined with other sensing functions such as magnetometer or gyroscope, see Chapter 2 or Chapter 9 for additional cautions about potential problems affecting these kinds of sensors.

vibration sensor

11

An **accelerometer** can measure some aspects of vibration. However, this entry deals primarily with mechanical devices (often described as *vibration switches*) and piezo-electric devices (often described as *vibration sensors*) that are solely intended to measure vibration.

A *vibrometer* measures vibration using a laser beam aimed at a reflective spot applied to a surface. It is usually a laboratory instrument, beyond the scope of this Encyclopedia.

OTHER RELATED COMPONENTS

- **accelerometer** (see Chapter 10)
- **tilt** sensor (see Chapter 8)
- **force** sensor (see Chapter 12)

What It Does

A vibration sensor responds to repetitive mechanical motion. Most versions contain two switch contacts that are normally open and will close if the sensor vibrates in its designed frequency range. In some sensors, the frequency range and sensitivity are manually adjustable.

Large sensors are used as automatic shutdown switches responding to excessive vibration in machinery, and may be capable of switching substantial currents (10A or higher). Smaller versions can shut down domestic appliances such as a washing machine that is seriously out of balance during a spin cycle.

A vibration sensor can be used as a simple user-input device in toys and games.

A shock sensor can detect abuse of a sensitive device, for example, by including the sensor and a data logger when the device is transported.

Schematic Symbols

Either of the symbols in Figure 11-1 may represent a piezoelectric or piezoresistive vibration sensor, but they also represent other piezo-based devices.

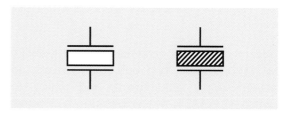

Figure 11-1 *Either of these symbols may represent a piezoelectric or piezoresistive device, including (but not limited to) vibration sensors that operate on this principle. The symbol on the left is more common.*

Variants

Vibration sensors use a wide variety of detection methods.

Pin-and-Spring

Probably the simplest type of sensor consists of a small, thin pin in the center of a miniature coil spring. The spring is anchored at its base, but its other end is free to vibrate. If the vibration reaches a sufficient amplitude, the spring touches the pin, completing a circuit between the two leads of the device.

An example is shown in Figure 11-2, where two identical sensors are shown, one of them cut open to reveal the gold-plated rod and spring inside. This sensor is rated for 10mA at up to 12VDC.

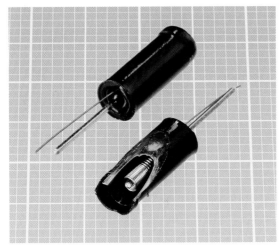

Figure 11-2 *When the spring in this sensor vibrates, it touches the pin that is centered in it. The background grid is in millimeters.*

Advantages of this system are low cost, ability to respond along two out of three axes, no power supply requirement, and ability to switch AC or DC. However, because the internal contact is extremely brief, it must be connected with a latching component of some type. A flip-flop could be used, or a 555 timer. The switch may also be connected with an input pin on a microcontroller, provided a pullup or pulldown resistor is used to prevent the input from floating when the switch is open.

Externally, the packaging of a pin-and-spring vibration sensor is almost indistinguishable

from a small **tilt** sensor that contains one or two spherical metal balls. The tilt sensor may respond to vibration, but only of a large amplitude and low frequency.

A miniature board containing a pin-and-spring vibration sensor, comparator, and trimmer potentiometer for sensitivity control is sold cheaply by the Chinese supplier Elecrow as their product SW-18015P, shown in Figure 11-3. Elecrow also offers a wide range of other low-cost sensors.

Figure 11-3 *A pin-and-spring vibration sensor mounted on a board with sensitivity control.*

Piezoelectric Strip

The LDT0-028K by Measurement Specialties is a section of piezoelectric polymer film laminated to a polyester substrate. The film is designed to be anchored at one end, allowing the other end to vibrate. An unweighted version and a weighted version are shown in Figure 11-4, each measuring about 13mm x 25mm. Addition of the weight alters the resonant frequency of the sensor.

Deflection of about 2mm is sufficient to generate a surprising 7VDC between the two leads. Larger deflections will generate higher voltages. The manufacturer suggests that direct connection to a CMOS component is possible. An op-amp may be used for signal conditioning.

A piezoelectric device of this type only generates voltage during the process of deflection. If the strip is held in a curved position, its output diminishes to zero.

This sensor has a resonant frequency around 170Hz when there is no weight attached to its free end.

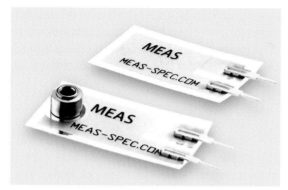

Figure 11-4 *Two versions of the LDT0-028K vibration sensor from Measurement Specialties, one with an added weight to lower its resonant frequency.*

Chip-Based Piezoelectric

The Murata PKGS series is an example of a surface-mount piezoelectric shock sensor. Measuring only about 1mm x 2mm x 4mm it has an analog output designed for connection through an op-amp. The manufacturer suggests application in a hard-disk drive to block read-write operations when vibration occurs. Similarly, these shock sensors may be used in CD-ROM or DVD drives. They may also be installed in cash dispensing machines to sound an alarm if vandalism occurs.

Toshiba's TB6078FUG is a similar product. Note that because these devices contain electronic components, they require a power supply (usually 3.3VDC to 5VDC) for operation.

"Mousetrap" Type

Some vibration switches rely on a simple system of leverage that is comparable to a mousetrap, in that a relatively small stimulus releases a strong spring. In Figure 11-5 the upper part of the figure shows the switch at rest, held in its position by a powerful spring and by the weight of a mass attached to a pivoted arm. In

the lower part of the figure, severe vertical vibration has caused the assembly to move up and down with sufficient energy to overcome the tension in the spring while the inertia of the mass has resisted the motion. Consequently the arm has moved past the position where the spring is aligned with the pivot, and the spring now acts to hold the arm against a snap-action switch. A system of this type is used in sensors on some power-station cooling towers, where the loss of a large fan blade can result in major vibration.

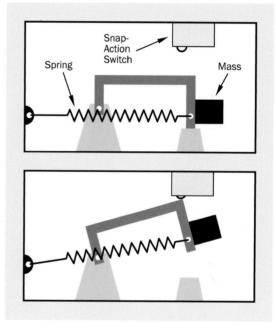

Figure 11-5 *A spring-loaded vibration sensor.*

Magnetic

Typically this method is used to detect excessive vibration in machines or other devices containing heavy rotating mechanical components. The sensor may be physically large, moving switch contacts that are designed to handle currents of 1A or significantly more.

In one system, a steel ball is retained by a permanent magnet that is barely powerful enough to prevent the ball from falling. Excessive low-frequency vibration will dislodge the ball, which falls and completes a circuit between two contacts. This activates a relay that powers down the piece of machinery that is vibrating.

The ball may be spaced a small distance from the magnet by a beveled nonmagnetic seat, and the seat may be movable with an external screw. This will adjust the sensitivity of the switch. After the switch has been triggered, it must be reset, which may entail using an external lever to raise the ball back to its location near the magnet.

Another magnetic system is shown in Figure 11-6, where a magnet on a vertical arm can be displaced by horizontal vibration, and an inertial mass on a spring-loaded horizontal rod can also dislodge the magnet in response to vibration along the other two axes.

Figure 11-6 *A magnetic vibration switch. See text for details.*

Mercury

A small *mercury switch* may be used as a vibration sensor, although this application is uncommon. See Chapter 8 for more information about mercury switches.

Values

Measurement of vibration is a complex science of special interest in mechanical design, especially in areas such as the powertrain and suspension geometry in an automobile. Only a few fundamentals will be summarized here.

Primary Variables

The four primary variables in vibration are frequency, displacement, velocity, and acceleration. Frequency describes how rapidly the vibration occurs; displacement describes how far the vibrating object moves in each direction; velocity describes how fast it moves during each cycle; and acceleration describes how rapidly the velocity changes during each cycle. Different types of sensors can be chosen for their responsiveness to each attribute.

Figure 11-7 shows the theoretical relationships between displacement, velocity, and acceleration plotted against the frequency of vibration. The y axis of this graph (the vertical axis) is often labeled "amplitude," but in reality it is being used to measure three different units, as shown. The curves indicate that if the velocity of vibration remains constant while frequency increases, acceleration must increase as a function of the frequency while displacement decreases. The acronym "rms" denotes that the values are measured as the "root mean square" of their fluctuations.

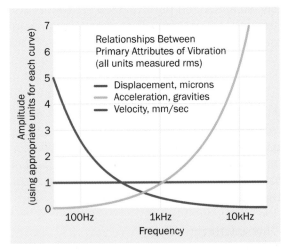

Figure 11-7 *Theoretical relationships among the primary attributes of vibration.*

Mechanical sensors or switches that respond to displacement are best suited to low frequencies, while piezoelectric sensors that are sensitive to acceleration are best suited to higher frequencies. Serious mechanical problems tend to result in low-frequency vibrations, while wear on bearings and in gear trains will be more likely to create high-frequency vibration.

Dynamic Attributes

Datasheets for small piezoelectric sensors may only show basic values such as the voltage range that is likely to be created.

Small mechanical sensors of the pin-and-spring type will be rated for maximum voltage and switching current. Typical values may be around 12V (AC or DC) and 10mA, indicating that the output from this type of sensor should be used with an op-amp, microcontroller, logic chip, solid-state relay, or other semiconductor with a high-impedance input.

Sensors that are manufactured for industrial applications will be rated for attributes such as measurable acceleration range (in gravities), temperature sensitivity, frequency response, resonant frequency, capacitance, and power requirement.

The sensitivity of piezoelectric sensors is usually expressed in mV/g. This type of sensor will require a comparator to process the tiny amount of current that it creates.

How to Use It

If a vibration sensor has an analog output requiring a comparator, the output from the comparator is likely to be of the open-collector type. This will require a pullup resistor of a value that provides an appropriate voltage for the next stage in the circuit. For more information about comparators, see Volume 2. For more information about the use of an open-collector output, see "3. Analog: Open Collector".

A *coupling capacitor* can remove the DC component from the comparator output, allowing only the frequency of vibration to pass

through. The choice of capacitor value will depend on the frequency.

When using a piezoelectric sensor with analog output, a 10M resistor may be installed across its two terminals to reduce voltage drift.

The primary challenge in getting a vibration sensor to work successfully will be matching it to the source of vibration. Manufacturers' datasheets for chip-sized sensors often provide very little information about the optimal values for their products. Peak performance will occur when the natural resonant frequency of a sensor is close to the frequency of vibration that it must detect. Trial and error may be necessary.

A sensor must be mounted appropriately. Most sensors are directional, at least to some extent, and many will not respond significantly to vibration at 90 degrees to their primary axis of sensitivity. Their performance will also diminish if they are placed too far from the vibration source, or if they are mounted on a flexible or yielding surface that will tend to absorb vibration.

While industrial vibration switches may be adjusted manually, the response of small devices designed for circuit-board mounting can only be tweaked using external components to filter out unwanted signals from the sensor.

What Can Go Wrong

Long Cable Runs

The output from a piezoelectric vibration sensor is primarily an AC signal, fluctuating at the frequency of the vibration. Long cabling, or inadequately shielded cabling, can potentially introduce capacitive effects that can degrade the sensor signal. This issue will only affect higher frequencies.

Interference

Sensor signals can also be affected by electromagnetic interference from power lines, transformers, and large motors. This is a significant issue, as a primary application for industrial vibration sensors is to measure vibration created by motors.

Correct Grounding

For large sensing equipment, grounding may be important to shield cables that transmit data. In an industrial environment, however, grounding is primarily motivated by safety considerations, and an electrical ground can carry unwanted interference. Ground loops may be created if there are multiple ground points. Ideally, a "ground tree" should be used, where there is only one primary grounding point, and grounds to equipment branch out from it.

Fatigue Failure

In installations where some vibration normally exists, cables should be anchored properly to minimize the risk of fatigue failures.

force sensor

12

A *load cell* or *load sensor* is generally intended to measure a static load, while a *force sensor* can respond dynamically. However, this semantic distinction is not always observed. This entry differentiates between load cells and force sensors but includes them both.

Traditional types of *hydraulic* and *pneumatic* load sensors are not electronic devices. They are outside the scope of this Encyclopedia.

Capacitive, *ultrasonic*, *magnetic*, *optical*, and *electrochemical* force sensors are relatively unusual, and are not included in this Encyclopedia.

A force sensor is occasionally described as a **pressure** sensor, but that term is ambiguous, as it is more often used in conjunction with fluids. This Encyclopedia assumes that a pressure sensor only measures gas or liquid pressure. See Chapter 17.

A **vibration** sensor reacts to rapidly changing forces, but usually cannot measure them accurately, and is simply triggered when vibration exceeds a threshold. See Chapter 11.

Impact sensors that measure the force of a collision are outside the scope of this Encyclopedia.

A sensor designed to respond to a single touch from a fingertip is considered a human-input device and is discussed in the **single touch** sensor entry. See Chapter 13.

OTHER RELATED COMPONENTS

- **vibration** sensor (see Chapter 11)
- **single touch** sensor (see Chapter 13)

What It Does

A **force** sensor measures physical force that is applied to it, either by a person or by an object. Many force sensors respond rapidly and can measure fluctuating forces.

A *load cell* or *load sensor* is usually intended to measure the static weight of an object.

Applications

In robots, force sensors can provide feedback to limit the grip of a mechanical hand. They can also provide haptic feedback for a surgeon performing robotic surgery. In the future, force sensors may find increasing application in agriculture, as mechanized handling of fruit and other foods requires a carefully controlled gripping force.

In medicine, the use of force sensors to evaluate muscle strength in hands or limbs can be important as an indicator of neurological problems or to monitor progress in occupational therapy. Thin force sensors can be installed in shoes to check the weight distribution of each foot. They can also be used for entertainment purposes, to light LEDs in sneakers.

A force sensor may respond to a single-touch user input. See Chapter 13 for more information about **single touch** sensors. Some video game controllers use resistive force sensors to measure the amount of pressure applied to a button. The PlayStation is an example. (Old PlayStation DualShock 2 controllers are a salvageable source of pressure-sensitive buttons.)

Load sensors are used to weigh industrial products, and are also used domestically in kitchen and bathroom scales.

A load sensor can also detect human presence —for example, in the passenger seat of an automobile, where an air bag must not deploy if a young child is present, or in a hospital, to monitor the number of times the patient gets out of bed.

Schematic Symbol

No specific schematic symbol is used for either a force sensor or a load sensor. If a force sensor uses a piezoelectric or piezoresistive element, it may be represented by the symbol shown in Figure 11-1, which is used for many piezo devices.

How It Works

Two methods of force measurement are commonly used: resistive and piezoelectric.

A *piezoelectric* force sensor uses a piezoelectric element, often consisting of a quartz crystal, to convert force to a small voltage that can be amplified. However, this type of sensor only responds to changes in force. If a constant load is applied, the output peaks quickly and then gradually diminishes to zero.

Resistive force sensors change their electrical resistance when force is applied. They include *metallic strain gauges* and *plastic-film* sensors in which two layers of conductive ink are pressed together.

In SI (standard international) units, the force needed to activate a sensor is measured in newtons, abbreviated with a capital letter N. A newton is defined as the force that would accelerate a mass of 1 kilogram at 1 meter per second each second. More practically, in the gravitational field at the surface of the Earth, 1N = about 100 grams of weight. There are about 28 grams in an ounce; thus 1N is slightly less than 4 ounces.

Strain Gauge

A strain gauge is often made from metallic foil applied to an insulating flexible backing. The backing is glued to a shaped piece of metal, usually steel or aluminum, which is designed to flex slightly under pressure and may be referred to as a *spring*, even though it is often one solid object. Its deflection will be related to the force imposed on it.

The maximum deformation of the spring under a strain gauge is usually 500 to 2,000 parts per million (ppm) when subjected to the maximum force that it is designed to measure. A change of 1ppm is referred to as a *microstrain* (abbreviated με).

The strain gauge has no polarity, and functions like a force-controlled **potentiometer** (see Volume 1). The ratio of the change in its resistance to the change in the strain that it experiences is called the *gauge factor*. For metal foil gauges, the gauge factor is usually around 2.0. This is an approximately linear relationship.

The most common type of foil pattern is shown in Figure 12-1. In the figure, if a stretching force is applied horizontally, the multiple thin sections of foil are slightly elongated, and their resistance increases within the limits of elasticity of the foil. This effect is multiplied by the number of sections. If the stretching force is applied vertically, the sections are merely separated slightly, and the result is negligible.

Wheatstone Bridge Circuits

The very small changes in resistance in a strain gauge must be amplified to be usable, and the first step is to use a Wheatstone bridge circuit. The simplest form of this circuit is shown in Figure 12-2.

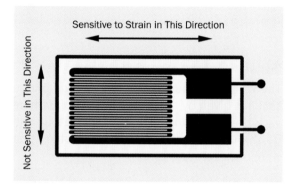

Figure 12-1 *The pattern of metallic foil used in a typical strain gauge.*

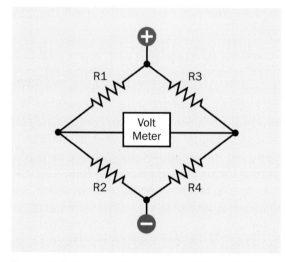

Figure 12-2 *A demonstration version of the basic Wheatstone bridge circuit that is often used to detect small changes in a resistance.*

Each pair of resistors (R1 + R2, and R3 + R4) functions as a voltage divider. If all the resistors have an exactly equal value, the voltage at the midpoint of each pair will be identical, and the volt meter at the center will have a zero reading. However, if the value of one resistor changes slightly, the meter will register the imbalance. This circuit is commonly used because of its sensitivity to small variations.

In Figure 12-3 two strain gauges have been substituted for resistors R3 and R4. The upper strain gauge has been mounted so that it experiences an increase in force at the same time that the lower strain gauge experiences a decrease, usually because one gauge is mounted on the top side of a flexing element while the other gauge is mounted on the underside.

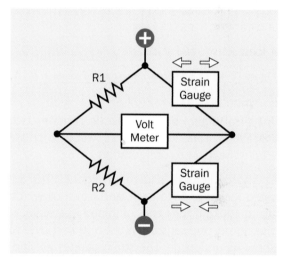

Figure 12-3 *Two strain gauges, oppositely oriented, can be used as resistances in the Wheatstone bridge circuit.*

Using two strain gauges in this way doubles the sensitivity of the Wheatstone bridge circuit. The configuration is known as a "half Wheatstone" force sensor, and will have three connecting wires. One will be black, one will be red, and the third will be a different color. The red and black wires are for connection to the power supply, as shown in the schematic, while the third wire is common and should be considered as an output.

If an additional two strain gauges are inserted in the Wheatstone bridge circuit, this is now a "full Wheatstone" force sensor (see Figure 12-4). Note, however, the diagonally symmetrical orientations of the strain gauges, necessary to multiply their effect.

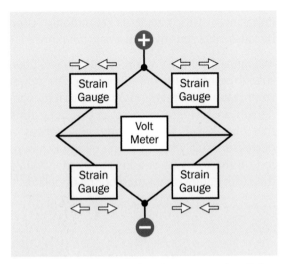

Figure 12-4 *Two additional strain gauges create a "full Wheatstone" circuit.*

A typical digital bathroom scale contains two half-Wheatstone force sensors, wired to create a full-Wheatstone configuration.

The force sensors shown in Figure 12-5 are rated for up to 50kg each, and can thus weigh up to 100kg if they are combined in a scale. In this figure, one sensor has been turned over to show its underside. The strain gauges are hidden in each sensor where the steel sections overlap.

Wheatstone Bridge Errors

Where more than one strain gauge is used in a Wheatstone bridge, they should ideally have identical performance. Since this is impossible as a result of manufacturing tolerances, some error correction is built into devices using multiple strain gauges.

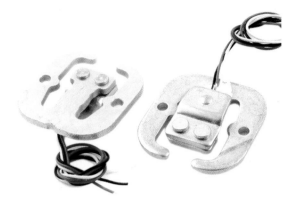

Figure 12-5 *Two force sensors suitable for a bathroom scale, each containing a pair of strain gauges that will flex oppositely under load.*

Strain-Gauge Amplification

The voltage output from a Wheatstone bridge circuit is given by the following formula, where V_{IN} is the supply voltage, V_{OUT} is the output, and R1 through R4 are the resistance values that were used in Figure 12-2.

$$V_{OUT} = [(R3/(R3+R4)) - (R2/(R1+R2))] * V_{IN}$$

The good news is that the output when a strain gauge is used will vary linearly with the load applied. The bad news is that it will be very small.

To amplify it, an op-amp such as the AD620 is often recommended. Using external resistors, its amplification factor can be adjusted from 1:1 to 10,000:1. Alternatively a chip such as the HX711 by Avia Semiconductor contains a 10-bit analog-to-digital converter and is specifically designed for use in weighing scales. Its digital output uses a very simple serial format. Sparkfun sells a breakout board incorporating this chip. See Figure 12-6.

Other Strain-Gauge Modules

Strain gauges are built into a variety of sensor modules. Figure 12-7 shows a selection, all of which are available from Sparkfun. Many more can be found online.

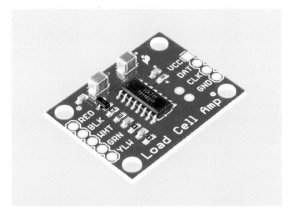

Figure 12-6 *Sparkfun HX711 breakout board for the HX711 amplifier chip, specifically designed for a full-Wheatstone array of strain gauges in a force sensor.*

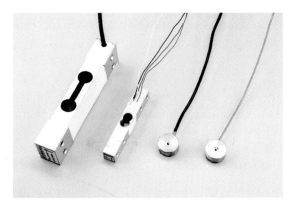

Figure 12-7 *A selection of load cells containing strain gauges, available from Sparkfun.*

Plastic-Film Force Sensors

Plastic-film resistive sensors contain two layers of conductive ink, sealed between two layers of thin, transparent plastic film. The resistance between the ink layers diminishes when they are pressed together. It may vary from as little as 30K when fully loaded to more than 1M when unloaded.

Like a strain gauge, this sensor has no polarity and requires no power supply.

Examples of this type of sensor are shown in Figures 12-8 and 12-9. Other shapes and sizes are available. Manufacturers include Tekscan, whose product is named FlexiForce; Alpha Electronic, in Taiwan; and Interlink Electronics.

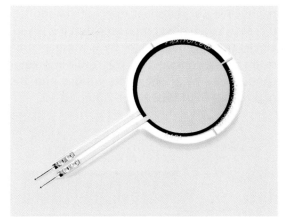

Figure 12-8 *A FlexiForce A401 resistive film sensor made by Tekscan, Inc. Its sensing area is 25.4mm in diameter (1 inch) and is rated to measure up to 111N (25lbs).*

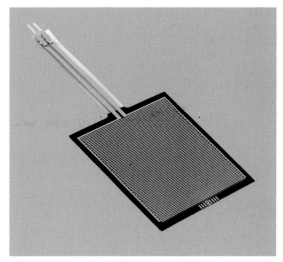

Figure 12-9 *An Interlink FSR406 resistive film sensor. Its sensing area is slightly less than 40mm square, and is rated to measure up to 20N (4.5lbs).*

Plastic-film sensors should not be confused with film-based piezoelectric vibration sensors, which are described in the **vibration** entry (see "Piezoelectric Strip"). Those sensors provide a transient output when they flex rapidly. The resistive sensors provide a stable output in response to a steady load.

Deformative Force Sensors

A sheet of natural or silicone-based rubber can be impregnated with conductive particles. The conductivity of the sheet may not change significantly when it is compressed, but if it is separated from a metal plate by a mesh of thin nylon fibers, compression will result in greater conductivity by pushing the rubber into the gaps between the fibers. See Figure 12-10.

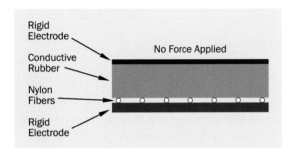

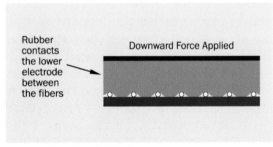

Figure 12-10 *In the top image, a flexible conductive layer is separated from an underlying rigid electrode by nylon mesh. In the lower image, a load applied to the flexible layer has forced more of it into contact with the electrode, reducing the resistance between them.*

Improvised Resistive Sensors

Polyethylene film impregnated with carbon particles is available under the brand name Velostat, owned by 3M. Although it was developed as an antistatic packaging material for semiconductors, it can be used to make a DIY force sensor. When the material is stretched, the embedded particles are more dispersed, and electrical resistance increases. When the material is compressed, its resistance is reduced.

Antistatic foam of the type used to package CMOS components can be used in the same way, although some types behave like memory foam, being slow to recover from pressure. The foam can be sandwiched between a pair of copper-plated circuit boards as electrodes.

How to Use It

Plastic-Film Resistive Force Sensors

The conductivity of a plastic-film sensor has an almost linear relationship with the force applied. In other words, if F is the force and I is the current:

$$I = k * F$$

where k is a constant determined by the characteristics of the materials used.

By Ohm's Law, R = V / I where V is the voltage drop across a resistance of value R. By substitution, using k * F instead of I:

$$R = V / (k * F)$$

Therefore, if a constant voltage is applied across the force sensor, the resistance of the sensor will be proportional to the reciprocal of the force (i.e., 1 / F). These relationships are shown in Figure 12-11.

For convenient measurement, it will be helpful if the resistance of the sensor can be converted to a voltage output that varies linearly with the force applied. To achieve this, a resistive force sensor of this type is customary amplified with an **op-amp** (see Volume 2). Using the schematic shown in Figure 12-12, the amplification ratio, A, is found from this formula:

$$A = 1 + (R2 / R1)$$

where R2 if the potentiometer, and R1 is the resistive force sensor. Therefore, the output from the op-amp should have an approximately linear relationship with the force applied to the sensor.

The capacitor in this circuit is added to suppress noise that the circuit might otherwise pick up.

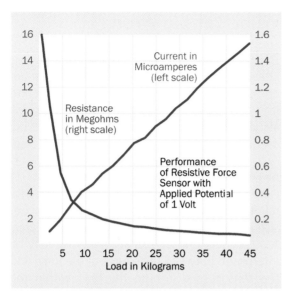

Figure 12-11 *The relationship between force, current, and resistance in a flexible resistive sensor, assuming a constant voltage of 1V is applied across it (derived from FlexiForce Sensors User Manual).*

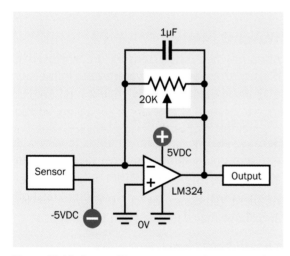

Figure 12-12 *An amplification circuit with an output that is approximately linear with force applied to a resistive film sensor.*

An alternative is to wire a resistive sensor in series with a capacitor, and connect the other side of the capacitor to a comparator that has adjustable feedback. The resistance of the sensor will determine how quickly the capacitor charges. However, because of the mathematics describing the charge rate, the output from the

comparator would not be linear with the force on the sensor. Also, provision would be necessary to discharge the capacitor intermittently.

Values

Film-Based Force Sensors for User Input

A very light pressure with a fingertip could be around 50g. A more defined finger-press would be 250g, and a heavy push with a finger would be around 1kg.

The Interlink range of flexible force sensors requires a minimum pressure of 0.2N, or about 20g. Similarly, Alpha products range from a minimum of 10g to 30g.

These specifications suggest that film-based force sensors may be used for one-touch user input, but the no-load resistance will be at least 1M, and more than 10M in some instances. A small amount of pressure is unlikely to reduce this electrical resistance much below 500K. Using an op-amp or comparator to detect that difference and convert it into a reliable on-off output may be vulnerable to noise and power-supply disturbance.

Another consideration is that film-based sensors provide no tactile feedback. For these reasons, and because film-based products are described by their manufacturers as "force sensors" rather than as "touch sensors," they are included in this entry rather than in Chapter 13. That said, they may be considered as an option for one-touch user input where they are appropriate—for instance, in games where players are likely to slap or hit a sensor vigorously.

Specifications for Film-Based Force Sensors

Durability of film-based sensors is excellent, with manufacturers claiming that performance is not degraded after 1 million applications of a 20kg load. Sensors have no polarity, will function using voltages from 1V to 15V in most cases, and have a response time of less than 5ms. They do not generate, and are not vulnerable to, electromagnetic interference.

Important attributes when evaluating film-type sensors are their limits of tolerance for applied force, and resistance values at each end of the range. Unfortunately these values are poorly documented for many sensors, or may be unspecified.

Maximum force may range from 20N to 440N, depending on the brand and model of sensor.

Electrical resistance is at least 1M when unloaded, and may be as high as 20M.

Accuracy ranges from plus-or-minus 2% to plus-or-minus 5% from one application of force to the next, depending on the model of sensor and the manufacturer. If force is not applied each time in exactly the same area of the sensor, results will vary. If a sensor is swapped with another sensor of the same type, sensor-to-sensor consistency will be uncertain. Therefore, film-based force sensors are not a good choice for applications where accuracy is important.

Sensors may be rated for 5% to 10% hysteresis.

The area of active detection may range from about 4mm wide (FlexiForce sensors) to more than 40mm wide (Interlink FSR-406).

Strain Gauges

Strain gauges are not sold as individual components by most electronics suppliers. They are sometimes available as surplus parts or from sites such as eBay, where specifications range from 100 ohms to 1K as the resistance when no load is applied.

A load sensor on which a strain gauge has been preinstalled is much easier to use, and will be plug-compatible with an appropriate amplifier chip as previously described.

What Can Go Wrong

Soldering Damage

The pins on plastic-film resistive sensors are embedded in thin plastic. Heat from a soldering iron can easily damage this plastic. Heat-sink alligator clips should be used while soldering, or the pins can be socketed instead of soldered.

Bad Load Distribution

Film-based sensors will not provide accurate readings if a load is imposed unevenly or inconsistently, or extends outside the detection area. A *puck* consisting of a small, rigid disc may be interposed between the source of the force and the sensor, to distribute the load within the maximum area. A puck may also be referred to as a *shim*.

Similarly, the sensor must be mounted on a flat, smooth surface, and if this is not available, a rigid plate should be interposed.

Water Damage

Although film-based sensors are enclosed in plastic, they are not waterproof. Immersion may cause the layers to delaminate.

Temperature Sensitivity

Because electrical resistance tends to vary with temperature, readings from resistive force sensors will vary with temperature.

Ambient temperatures of 70 degrees Celsius and above may damage a film-based sensor.

Leads Too Long

Although film-based sensors are supplied with a variety of lead lengths enclosed in the laminated layers of flexible plastic, the leads may be too long for a particular application. If they are trimmed, wires cannot be attached with solder, as it will melt the plastic. Conductive epoxy should be used.

single touch sensor

13

This entry only describes *capacitive* touch sensors. A *conductive* sensor, which uses the fingertip to complete a circuit between two exposed contacts, is not very common, and is not included here.

The type of touch sensor described in this entry requires no physical pressure for activation. It should not be confused with a resistive or piezoelectric **force** sensor that requires pressure. See Chapter 12.

An integrated circuit chip that processes a signal from a *touch pad* is often described as a *touch sensor*, even though it does not contain a sensing element. This entry describes it as a "touch sensor chip" to eliminate ambiguity, and refers to touch input elements as "touch pads."

Touch pads that contain *tactile switches* or *membrane switches* are described in the entry discussing **switches** in Volume 1. All types of switches are described in that volume, with the exception of a *reed switch*, which is magnetically activated and is therefore categorized as a sensor.

Capacitive touch sensors are sometimes referred to as *capacitive proximity sensors*, because they sense the proximity of a human fingertip. In this Encyclopedia, and in most other sources, a **proximity** sensor measures distance, not touch. See Chapter 5.

A *capacitive displacement sensor* employs the same principle as a capacitive touch sensor, but is used to detect the position of an object, not for human input.

OTHER RELATED COMPONENTS

- **force** sensor (see Chapter 12)
- **touch screen** (see Chapter 14)

What It Does

A *touch pad* detects the presence of a human fingertip (or other part of the body) and signals an integrated circuit chip, which is very often termed a *touch sensor*, even though it does not contain a sensing element itself. The chip creates an output to signify that human touch has been recognized.

A keypad of the type found on microwave ovens may appear similar to an array of touch pads, but is more likely to contain *membrane switches* or *tactile switches*, which are described with other forms of switches in Volume 1. The type of touch pad described in this entry requires no physical force and contains no parts that move or flex when pressed.

A modern **touch screen** is usually a capacitive device and can be thought of as an array of touch pads. See Chapter 14.

Applications

Capacitive touch sensors have become common as their cost has fallen relative to simpler components that respond to being pressed.

A touch sensor can be used to start or stop a process, or to power up or power down a device. Multiple sensors may be found wherever user input of a few alphanumeric characters is required. Because touch pads can be completely sealed, they are useful where hygiene is important.

Specific applications include the activation of a backlight in a handheld device, wake-up from standby, ear detection in a cellular telephone, control of medical devices, and activation of interior lighting in some automobiles.

The absence of moving parts or electrical contacts means that a touch pad is more reliable than any type of electromechanical switch. A disadvantage is that it provides no tactile feedback, and therefore will require a visual or audible confirmation when it responds to input. Lack of tactile feedback makes touch pads unsuitable for computer keyboards and other key-entry devices where rapid typing is required.

When the capacitive elements of a touch pad are transparent, it can be mounted in front of a screen.

Schematic Symbols

Either of the schematic symbols in Figure 13-1 may sometimes be used to represent a touch sensor, but not on a consistent basis.

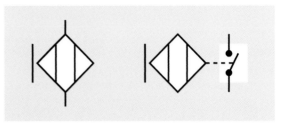

Figure 13-1 *Two possible schematic symbols that may represent a touch sensor.*

How It Works

The two plates in a capacitor are separated by an insulator known as the *dielectric*. Paper, plastic, glass, air, and other insulators can serve this purpose. Although no electrical connection exists between the plates, AC passes through the dielectric as a field effect.

Electrical capacitance exists between any pair of electrical conductors. The human body has high electrical resistance but still is electrically conductive, and therefore has capacitance with other conductive objects.

A touch pad can function as one side of a capacitor, with a fingertip functioning as the other side. In this mode, AC can pass from the touch pad and through the human body to ground. The current is very small, but fluctuations can be detected by an appropriately designed integrated circuit chip, or by a microcontroller.

The precise characteristics of a dielectric will affect the performance of a capacitor to some extent, but will not prevent it from working. Therefore a touch pad can function even when it is shielded behind a protective layer of glass or plastic, as is often the case.

A touch sensor chip generates pulses of low-voltage AC and sends them to a touch pad. The chip detects any variation in the current through the touch pad, indicating the presence of a fingertip. Where an input occurs, the chip changes its output, which usually requires a microcontroller for processing.

How to Use It

Touch sensor chips are available in as many as 40 different formats and configurations. All of them are surface-mount. For breadboarding, an experimenter can use breakout boards that have sensor chips installed. Figure 13-2 shows a product from Adafruit that is capable of addressing 12 touch pads. Its output is accessible from a microcontroller via the I2C protocol. For additional details about protocols such as I2C, see Appendix A.

Figure 13-2 *A capacitive touch sensor chip on a breakout board from Adafruit.*

Similar breakout boards are available from Sparkfun, and from large online vendors such as *http://www.mouser.com* (where they are categorized as *development tools*).

While most touch sensor chips require a microcontroller, a few are available with the same number of output pins as touch-pad input pins, and each output pin will transition between logic-high and logic-low when an input on the corresponding pin is detected. Another breakout board from Adafruit, the AT42QT1070, uses this simple system.

A library (*http://bit.ly/1WJ9kE8*) exists for Arduino that enables two pins to sense touch on a piece of aluminum foil.

It can also work with conductive ink or paint (*http://bit.ly/1WJ9jQM*).

Obtaining Touch Pads

Sensor chips are widely available as components, and are very inexpensive. On the other hand, touch pads are not common as components, probably because a touch pad is usually created as a pattern of copper traces etched onto a circuit board by a device manufacturer.

Touch pads from hobby-electronics sources usually include touch sensor chips. Sparkfun offers a 12-key keypad on this basis, and also a 9-key keypad designed as a "touch shield" for use with the Arduino. Both of the Sparkfun products are shown in Figure 13-3. They include the same MPR121 touch sensor chip as the breakout board from Adafruit shown in Figure 13-2, and require an I2C connection with a microcontroller.

Figure 13-3 *Two capacitive keypads from Sparkfun, the one on the right designed as an Arduino shield.*

Because a capacitive touch pad is usually mounted inside an enclosure, the appearance of the bare touch pad as a component is unimportant. The outside of the enclosure can be printed with a design showing key outlines.

Individual Touch Pad

Adafruit sells the AT42QT1010 touch pad that emulates a momentary switch. Its output transitions from logic-low to logic-high when a finger presence is detected, and transitions back to logic-low when the finger is removed.

An alternative, shown in Figure 13-4, toggles between a logic-high and logic-low output and latches in each state with each single key press.

Both of these keypads contain sensor chips to generate the output.

Figure 13-4 *The output from the AT42QT1012 sensor chip on this touch pad from Adafruit toggles between logic-high and logic-low each time the pad is touched.*

Wheels and Strips

A touch wheel uses a circular pattern of conductive traces, often referred to as *electrodes*, to receive finger input. A simple configuration is shown in Figure 13-5. The traces interlock without making contact with each other, so that moving the finger in a circular motion creates a capacitive input that rises and falls sequentially on each element. In the figure, three sections are used, each colored differently here for purposes of clarity. Other touch wheels may contain more sections.

Firmware that is designed to interact with wheel-shaped touch pads will generally assume that two electrodes are receiving input at one time. The firmware attempts to calculate the position and motion of the finger by assessing the relative capacitance of adjacent segments. Ideally the capacitance values should vary linearly and complementarily; that is, from a 50-50 value at a midpoint between two segments, the progression should change to 60-40, 70-30, and so on, as the finger moves around the wheel.

A *touch potentiometer* consists of multiple touch pads, usually arrayed in a straight line. This may be described as a *touch strip*. An example is made by GHI Electronics and sold by

Robot Shop as their L12 Capacitive Touch Module, shown in Figure 13-6.

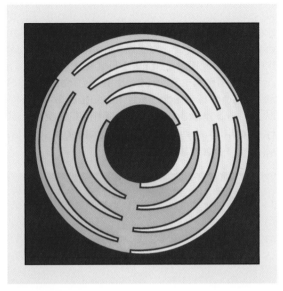

Figure 13-5 *A touch wheel created by copper traces on a circuit board (green). Each section is colored differently, for purposes of clarity.*

Figure 13-6 *Multiple capacitive touch pads arrayed as a strip, made by GHI Electronics.*

Design Considerations

A simple touch pad is often surrounded by additional copper that is grounded and may be described as a *shield* or a *guard*. The capacitance of a fingertip (spaced above the electrode by a layer of plastic or glass that functions as the dielectric) interferes with the field between the electrode and the shield.

The underside of the circuit board is often plated and grounded to protect the touch pad from electromagnetic interference. This, too, can be referred to as a shield or a guard. The ground plating can be in a hatched pattern to reduce its capacitance with the electrode above.

Layout of a circuit involving a single touch sensor chip can affect its performance significantly. A touch sensing device may have to be "tuned" to detect finger presence reliably.

A long trace between the touch sensor and an electrode will tend to pick up noise and will increase capacitance.

The distance between adjacent traces from multiple touch pads must be maximized to reduce capacitance between them. If output from a sensor chip uses the I2C or SPI digital protocol, any trace carrying that digital signal should be at least 4mm from input traces. If they cross, they should be at 90 degrees to each other.

Electrodes should not be shaped to resemble numbers or other characters printed above them. A single basic electrode should be circular.

What Can Go Wrong

Insensitive to Gloves

Gloves are a challenge for touch-sensor design, as they alter the dielectric and the distance between the electrode and the finger. Capacitive touch sensors may not work at all with some types of gloves. However, special gloves containing metallic threads are available.

Stylus Issues

A nonconductive stylus cannot activate a touch pad.

Conductive Ink

Ink that prints the shapes of touch pads on the exterior of a device should be nonconductive.

touch screen

14

The term **touch screen** is written as two words in this Encyclopedia. In many sources, the words are concatenated as "touchscreen."

The two words are hyphenated here when they function as an adjective, but not otherwise. In manufacturers' datasheets where the two words are used, they are not usually hyphenated.

OTHER RELATED COMPONENTS

- **single touch** sensor (see Chapter 13)
- **force** sensor (see Chapter 12)

What It Does

A touch screen is a video display with embedded touch sensing. The screen reports the position of the touch, and is used as a pointing device as an alternative to a mouse or trackpad. Some touch screens report pressure as well as position.

Touch screens are widely used in smartphones and tablets, and also in some laptop computers. Smaller, simpler touch screens may be found in office equipment such as photocopiers.

Schematic Symbol

No specific schematic symbol is used to represent a touch screen.

Variants

Early designs used infrared LEDs recessed into the edges of a frame around a screen. A matching photodiode picked up the focused beam from each LED. The presence of a fingertip was detected when it interrupted one or more of the beams. This system was not capable of high resolution, but was adequate for detecting user input at predefined locations.

Most touch screens currently are either resistive or capacitive.

Resistive Sensing

A resistive touch screen consists of two transparent layers that can be installed over a separate video display.

Each of the layers has uniform electrical resistance. Pressure from a fingertip on the outer layer (which we can refer to as layer 1) forces it to make contact at a point with the inner layer (referred to here as layer 2).

Two vertical electrodes connect with layer 1 along its left and right edges. Two horizontal electrodes connect with layer 2 along its top and bottom edges. When voltage is applied between the vertical electrodes on layer 1, the layer acts as a horizontal voltage divider. The voltage at the point where the layer is being pressed is applied to layer 2, and can be read from either of the electrodes on layer 2, so long as the metering has a much higher impedance than that of the layer. The voltage is decoded as a value for the horizontal position on layer 1.

An external switching device now repeats the procedure, except that it applies voltage to layer 2, and reads it from layer 1 to supply a value for the vertical position on layer 2. The sequence is illustrated by the top and bottom sections of Figure 14-1.

Because only four connections are necessary, this is referred to as a *four-wire* resistive touch screen. Five-wire variants exist, but are less common, and are not included in this entry.

Advantages of a resistive screen include:

- Simplicity. Only four connections are necessary, and the layers of the screen do not have to be subdivided into separate conductors.

- Low cost, relative to capacitive touch screens.

- Will respond equally well if a user wears gloves or uses a stylus.

Disadvantages of a resistive screen include:

- Some resistive versions require a stylus input instead of finger pressure.

- Resistive screens only respond to a one-location input. Two-finger gestures are not supported.

- Contact bounce occurs when the flexible layer is pressed against the underlying layer, and voltage spikes may be associated with switching power to the screen. To address this issue, firmware in a microcontroller may have to take a median value from several rapid readings.

- The flexible membrane is vulnerable to damage from sharp objects.

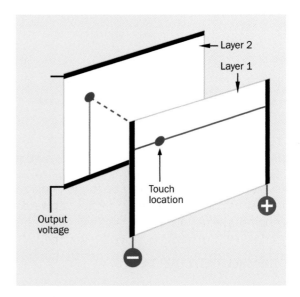

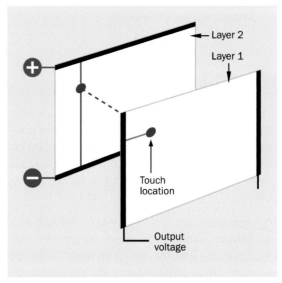

Figure 14-1 *Sections of a resistive touch screen are shown displaced for clarity. In reality they would be separated by a very small gap, allowing one section to make contact with the other in response to finger pressure.*

Capacitive Sensing

A capacitive touch screen can consist of an array of **single touch** sensors printed onto a glass panel as vertical and horizontal lines of transparent conductive ink.

Alternatively, a small capacitive screen can measure the tiny amount of current drained by a fingertip from four sources located at corners of the screen.

For more information about capacitive touch sensing, see the entry on touch sensors in Chapter 13.

Screens Available as Components

A wide variety of screens can be found in diagonal sizes ranging from 2 inches upward. Onboard electronics can be suitable for connection with a microcontroller using the I2C and SPI protocols, or USB. Different screen resolutions are available.

An example of a 3.5-inch touch screen with 320x200 resolution, mounted on a breakout board that can be used with a breadboard and an Arduino, is shown in Figure 14-2.

Figure 14-2 *An Arduino-compatible touch screen mounted on a breakout board, available from Adafruit.*

A 7-inch resistive touch screen that can be mounted on a separate video display is shown in Figure 14-3. It can be used with the STMPE610 controller chip, which converts resistive screen values into digital coordinates and can be accessed by a microcontroller over both SPI and I2C. This surface-mount chip is available on a breakout board.

Figure 14-3 *This resistive touch screen is intended for use as a layer applied to a 7-inch video display. The screen is available from Adafruit.*

When choosing a touch screen as a component for a DIY project, the availability of microcontroller code libraries for reading and refreshing the display is an important consideration.

liquid level sensor

15

Level indicators that contain no electronic components are not included in this entry.

Specialized industrial level-sensing equipment is generally outside the scope of this Encyclopedia. This entry discusses small-scale, lower-cost sensors.

Liquid volume can be assessed by measuring liquid pressure at the bottom of a reservoir. Sensors for this purpose are discussed in the entry describing **gas/liquid pressure** sensors. See Chapter 17.

OTHER RELATED COMPONENTS

- **liquid flow rate** sensor (see Chapter 16)

- **gas/liquid pressure** sensor (see Chapter 17)

What It Does

Measuring the volume of liquid in a storage vessel or reservoir is such a fundamental task, countless methods have been devised, of which only the simplest and most common will be discussed here.

A liquid level sensor can have a binary output, meaning that it signals when the volume rises above or falls below a level that can be preset or reset. Often the sensor will be connected to a pump or valve that maintains a relatively constant volume in a container.

Alternatively a sensor can indicate the actual stored volume, either with an analog output or in digital increments.

Schematic Symbols

Three variants of a schematic symbol for a simple liquid level sensor are shown in Figure 15-1. They are not always used, however, and a sensor may be shown simply as an annotated switch.

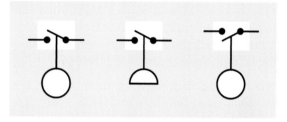

Figure 15-1 *Three variants of a schematic symbol to represent a liquid level sensor. The rightmost symbol indicates that a rising level closes, rather than opens, a switch.*

Applications

The fuel gauge in a vehicle is one of the most commonly encountered applications of a liquid level sensor. The water tank in a recreational vehicle or boat may use similar electronics. In industry, the choice of a sensor will be influenced by the type of liquid that is being stored, the desired accuracy, the temperature range, and whether the storage tank is sealed or open to atmospheric pressure.

How It Works

Desirable attributes of any liquid-level sensor include resistance to vibration, some damping to average out fluctuations caused by turbulence or sloshing in the liquid, resistance to chemical reactions with the liquid, and few moving parts that may require maintenance if the sensor is inside a sealed tank. Desirable attributes of an analog float sensor include a linear response, and some hysteresis if the application requires it.

This entry compares a variety of sensing strategies.

Binary-Output Float Sensor

The term "binary output" is used here to describe an output that only has two states (on and off, or logic-high and logic-low). The simplest type of liquid-level sensor with a binary output consists of a donut-shaped float that contains a permanent magnet and is free to slide vertically up and down a sealed tube containing a reed switch. The tube is supported on a bracket that can be mounted on the wall or the lid of the vessel containing the liquid.

The tube and float must be nonmagnetic, and the float must have a significantly lower specific gravity than the liquid that is being used ("significantly" because the float requires sufficient buoyancy to carry the weight of the magnet and overcome any friction between itself and the tube). A diagram illustrating this configuration is shown in Figure 15-2.

To change the level setting of the sensor, the bracket may be mounted on a screw thread to adjust its vertical position.

The reed switch can be normally open or normally closed, as needed to respond to a rising or falling liquid level. For basic information about reed switches, see "Reed Switch". Additional, detailed information about reed switches is included in the book *Make: More Electronics*.

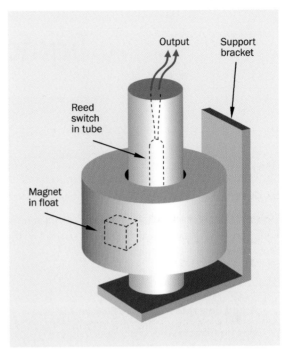

Figure 15-2 *The principal parts of a basic binary-output float sensor.*

For increased reliability, a Hall-effect sensor could be substituted for a reed switch. See "Hall-Effect Sensor" for general information about Hall-effect sensors.

Another binary-output float sensor is shown in Figure 15-3. This is a sealed plastic capsule containing a snap-action switch and a steel ball. The cable is attached to the underside of the top of a tank, and the capsule dangles into liquid in the tank. A separate weight (not shown) has a hole in the middle, and is threaded over the wire. The weight keeps the wire approximately in a vertical position as it dangles into the tank.

Figure 15-3 *An air-filled float that switches external power depending on its orientation.*

Figure 15-4 shows the components inside the float. When the liquid level in the tank falls, the float adopts the position shown on the left in the figure. The ball drops against a lever that closes a snap-action switch, which starts an external pump to replenish the tank. As the liquid level rises, the buoyancy of the air-filled float changes its orientation to that shown on the right in Figure 15-4. The ball drops and the switch opens, stopping the pump.

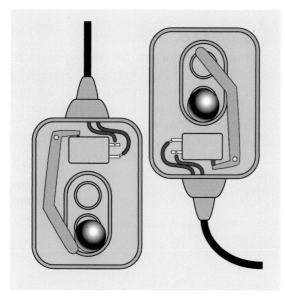

Figure 15-4 *Internal components of the float shown in the previous figure.*

Two circular indents on the inner surface of the plastic capsule prevent the ball from rolling erratically if there is turbulence in the liquid. They also provide some hysteresis.

Analog-Output Float Sensor

The simplest type of liquid-level sensor with an analog output consists of a float on an arm attached to a potentiometer, as shown in Figure 15-5. This very basic design was used in fuel tanks in vehicles for many decades. Disadvantages include a nonlinear response and the limited life expectancy characteristic of potentiometers. Some compensation for the nonlinear response can be made by using an analog fuel gauge with a nonlinear scale.

For more information about potentiometers, see "Arc-Segment Rotary Potentiometer".

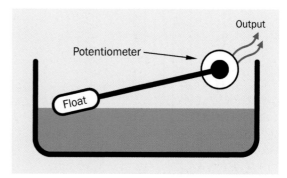

Figure 15-5 *A basic float sensor with an analog output.*

Incremental-Output Float Sensor

A schematic diagram for a simple float sensor with incremental output is shown in Figure 15-6. A magnet embedded in a donut-shaped float, similar to that shown in Figure 15-2, interacts with a sequence of reed switches installed in the central tube. The switches are spaced at equal intervals and apply power between resistors of equal value wired in series. This system has been used in motorcycle and automobile fuel tanks, where the switches may be enclosed in a (nonmagnetic) stainless-steel tube. The accuracy is limited by the number of reed switches.

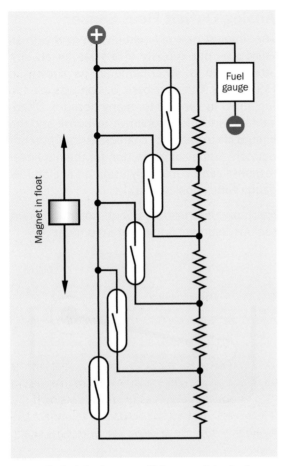

Figure 15-6 *A float sensor with incremental output.*

Displacement Level Sensors

If a heavy object, described as the *displacer*, is suspended in liquid, the effective weight of the object diminishes as the liquid rises up around it. This occurs because according to Archimedes' Principle, the upward buoyant force is equal to the weight of liquid that the object displaces. The displacer is suspended from a load sensor that measures its weight. Analog output from the sensor will be approximately linear with liquid level.

For a simplified diagram showing a displacement level sensor using this concept, see Figure 15-7.

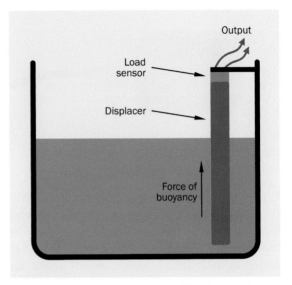

Figure 15-7 *A displacement sensor. The displacer is heavier than the liquid around it, but its effective weight diminishes as the liquid rises.*

For more information on load sensors, see Chapter 12.

Ultrasonic Level Sensors

An ultrasonic sensor can be used to measure the level of liquid in a reservoir, as shown in Figure 15-8. For more information about this type of sensor, see the entry discussing proximity sensors in Chapter 5. A disadvantage of using ultrasound for liquid level sensing is that the speed of sound will be affected by any vapor given off by a volatile liquid.

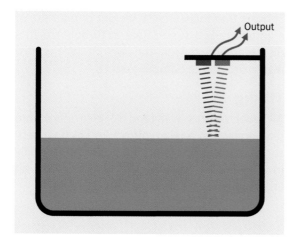

Figure 15-8 *An ultrasonic proximity sensor can measure the level of liquid in a reservoir.*

Reservoir Weight

The weight of a reservoir can be measured to assess the volume of liquid in it. This can be done by mounting the reservoir on load sensors. However, pipes leading to and from the reservoir must be designed so that they do not add or subtract any significant weight. Figure 15-9 suggests an arrangement, although the outflow will still change the weight to some extent depending on the amount of suction that is applied. For more information on load sensors, see Chapter 12.

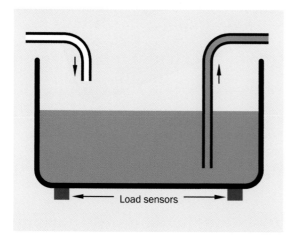

Figure 15-9 *Using load sensors to assess the weight of liquid in a reservoir requires that the weight of plumbing should not be imposed on the structure of the reservoir.*

Pressure Sensing

A differential pressure sensor can be added to a pipe near the bottom of a reservoir. The sensor measures the difference between the liquid pressure and ambient air pressure. See Figure 15-10.

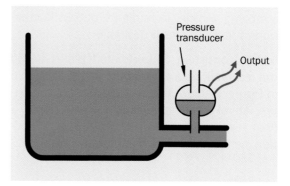

Figure 15-10 *A pressure sensor can assess the volume of liquid in a reservoir.*

This arrangement assumes that the reservoir is vented so that air pressure above the liquid level is equal to air pressure at the sensor. If the reservoir is not vented, a pipe must connect the reference port on the sensor with the space above the liquid.

The reservoir must have straight, vertical sides for the pressure to be directly proportional to the liquid volume.

Liquid volume can also be assessed by measuring the pressure inside a container, near the bottom. A submersible pressure sensor can be used, typically consisting of a watertight capsule fitted with a diaphragm that connects with an internal strain gauge. The sensor is lowered on a cable that also contains an air line. Because pressure in a liquid is affected by atmospheric pressure above the surface of the liquid, the sensor requires an air line so that its measurements are relative to the outside air.

Submersible pressure sensors are useful where access is limited—for example, when measuring fluctuations in an open-air municipal water reservoir.

What Can Go Wrong

Turbulence

Surface turbulence consisting of ripples, waves, or sloshing of liquid will tend to occur when a reservoir is refilled rapidly or is subjected to lateral movements as a result of being mounted in a moving vehicle. To minimize output fluctuations, some *damping* is desirable.

Baffles consisting of perforated plates inside the reservoir are a common strategy, as shown in Figure 15-11. In the upper section of the figure, lateral acceleration causes submersion of a float sensor in the reservoir. In the lower section, perforated baffles minimize the problem.

Sensors that measure the weight or pressure of a liquid are less susceptible to turbulence. In a displacement sensor, the weight of the displacer provides a damping effect.

Tilting

All level sensors will tend to give inaccurate readings when a reservoir is tilted. A float sensor will be affected less if it is mounted centrally in a reservoir, because the reservoir will tilt around the sensor, as shown in Figure 15-12.

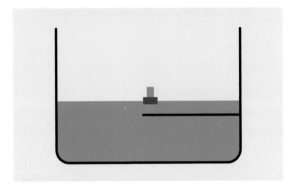

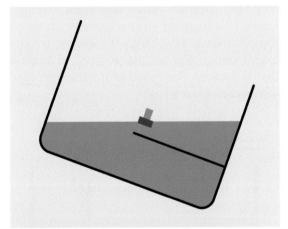

Figure 15-12 *If a float sensor is mounted centrally in a reservoir, it will be significantly less affected if the reservoir tilts.*

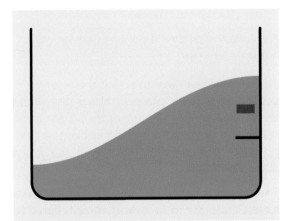

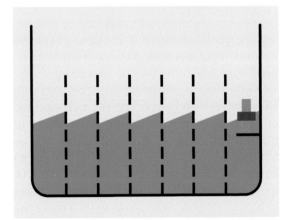

Figure 15-11 *Insertion of perforated baffles in a reservoir minimizes the sloshing that otherwise tends to occur when the reservoir is subject to lateral motion.*

liquid flow rate sensor

16

Flow rate sensors that contain no electronic components are not included in this entry.

Some methods of liquid flow sensing can also be applied to gases, but sensors are usually designed for one application or the other. Therefore, gas flow sensors have their own entry. See **Chapter 19**.

Many liquid flow rate sensors are large devices designed for industrial applications. This entry focuses on lower-cost solid-state sensors.

OTHER RELATED COMPONENTS

- **liquid level** sensor (see Chapter 15)
- **gas/liquid pressure** sensor (see Chapter 17)
- **gas flow rate** sensor (see Chapter 19)

What It Does

A liquid flow rate sensor measures the rate at which liquid flows past or through the device. A water meter is an example of a flow rate sensor.

A sensor may have a binary output, meaning that it signals when flow stops or starts, or if its rate rises above or falls below a level that can be preset or reset. However, most flow meters have an analog output that varies with the volume per unit of time.

Measuring the flow rate of a liquid can be challenging if the viscosity is very high, the liquid is chemically reactive, or the rate is very low. Such factors may require specialized equipment that is outside the scope of this Encyclopedia.

This entry compares the most popular sensing strategies.

Schematic Symbols

Many specialized symbols are used in flow diagrams to represent pumps, valves, and sensors. Typically they involve a single letter or an X in a circle. These symbols are not generally found in electronic schematics, and therefore they are not included here.

Paddlewheel Liquid Flow Rate Sensors

The simplest and most common liquid flow rate sensor uses a *paddlewheel*, also referred to as a *rotor*, that is mounted with its axis of rotation at 90 degrees to the direction of liquid flow. An example is the Koolance INS-FM16 shown in Figure 16-1. This sensor is intended for use in an aftermarket cooling system for the CPU in an overclocked desktop computer, but can be used in any system where the rate of flow ranges from 0.5 to 15 liters per minute. The paddlewheel has a pair of magnets mounted in it, activating a *reed switch* that is mounted in a sealed enclosure beneath the wheel. (For more information regarding reed switches, see "Reed Switch" in the entry describing **object presence** sensors.)

Figure 16-1 *A low-cost, simple paddlewheel sensor designed for rates of 0.5 to 15 liters per minute. The background grid is in millimeters.*

While the reed switch in the Koolance flow sensor inevitably suffers from contact bounce, it has the advantage of simplicity and can be used in conjunction with appropriate hardware or microcontroller code to debounce the pulse stream.

As in any device with a rotating part, friction and wear will afflict a paddlewheel sensor, especially because the bearings are often in the chamber through which the liquid passes. This eliminates the possibility for roller bearings. Typically a plain bearing is used, consisting of a pin that engages in a hole in the casing. Friction wears the bearing surface, creating a larger gap, which allows the rotor to vibrate or bounce instead of spinning smoothly. Flow resistance increases and accuracy is degraded.

In modern designs, the mass of the rotor is minimized to reduce the friction. Also, if the shaft is horizontal (but still at 90 degrees to the direction of the flow), buoyancy of the rotor in the liquid can reduce friction still further, as suggested in Figure 16-2. Ideally, the density of

the rotor and the density of the liquid will be the same.

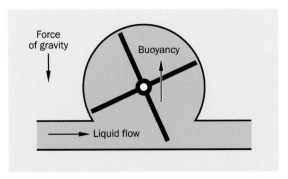

Figure 16-2 *In this configuration, friction on the rotor bearings is mitigated by taking advantage of the buoyancy of a low-density rotor in the liquid that passes through.*

The U-shaped flow path in the Koolance sensor in Figure 16-1 maximizes the responsiveness of the rotor, but an inline path is more common. An example is shown in Figure 16-3.

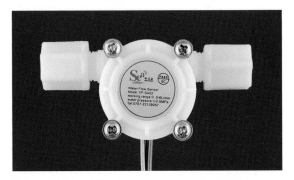

Figure 16-3 *An inline flow sensor rated for 3 to 6 liters per minute.*

Turbine Flow Rate Sensors

In a turbine-type sensor, two or more spiral blades are attached to a hub that rotates around an axis in line with the liquid flow, as shown in Figure 16-4. A magnet in each spiral blade triggers a reed switch or Hall-effect sensor mounted in a bracket that suspends the turbine from the interior walls of the tube.

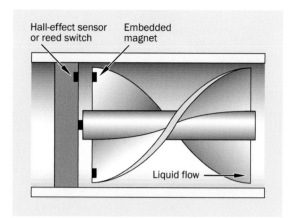

Figure 16-4 *Simplified view of a turbine flow rate sensor mounted inside a tube.*

The support bracket often consists of four struts, which impose resistance to liquid flow. The bearings suffer from the same kinds of problems as the bearings in a paddlewheel sensor, and must withstand additional load resulting from the inline force exerted by flow. Overall, while the turbine type of sensor is popular in laboratory equipment, it has disadvantages that are not shared by the bulkier paddlewheel type.

Limitations of Paddlewheels and Turbines

Both the paddlewheel and turbine types of sensors require a minimum flow to overcome the friction in their bearings. Below this minimum, liquid will find its way around the rotor without turning it. Even when the flow exceeds the minimum, response of the rotor is likely to be nonlinear as a result of turbulence and other factors.

Above a limit stated by the manufacturer, turbulence increases to the point where output from the sensor is no longer meaningful. Wear on the bearing will also increase with flow rate.

A significant problem for these types of sensors is that they do not respond well to sudden variations in flow. The paddlewheel, in particular, has inertia as a function of the diameter of the rotor, and will take some time to spin up in

response to an increase in flow. Conversely, when the flow diminishes, the paddlewheel will tend to overrun.

The viscosity of the liquid passing through a paddlewheel or turbine sensor will have a very significant effect on its performance.

Thermal Mass Liquid Flow Rate Sensor

The thermal-mass system is commonly used when volumes are extremely low. The system is illustrated in Figure 16-5. A tube containing liquid is fabricated from a heat-conductive metal such as aluminum. It is enclosed in a larger tube, and the gap between them is filled with thermal insulation. A temperature sensor such as a thermistor measures the temperature of liquid entering the system. A second sensor, combined with a small resistive heater in the form of a coil around the tube, is placed downstream. Liquid passing through the tube will tend to remove heat more effectively at higher flow rates, and the difference in temperature between the two sensors is a logarithmic function of the flow rate.

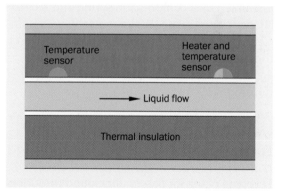

Figure 16-5 *In this type of low-flow sensor, the temperature differential between the two sensors is a logarithmic function of the flow rate.*

Variants of this system use slightly different tube configurations and sensor placement, but the principle is the same. Its advantages include the lack of any moving parts, and the avoidance of any probes intruding into the liquid, which is desirable in biochemical and medical applications.

The same principle is applied in many **gas flow rate** sensors. See "Mass Flow Rate Sensing".

Sliding Sleeve Liquid Flow Switch

This sensor is used in some domestic systems, where flow-activated water heating is required. A vertical section of brass (nonmagnetic) water pipe contains a sliding inner sleeve that incorporates a magnet. An external reed switch is activated when the sleeve is moved by water flowing through it. When the flow stops, the sleeve is returned to its rest position by the force of gravity.

Sliding Plunger Liquid Flow Switch

Figure 16-6 shows an exploded view of a similar device, using a plastic plunger that slides inside a nylon plumbing fixture designed for 3/4-inch pipe. The plunger contains a magnet and is restrained by a compression spring and a circular perforated plate. When water flow is sufficient to overcome the resistance of the spring, the plunger slides far enough for the magnet to activate a reed switch sealed into the external housing.

Ultrasonic Liquid Flow Rate Sensor

This type of sensor passes ultrasound through a liquid in a pipe. The speed of sound through the liquid is affected by the flow rate, and external electronics translate this lag time into a value for volume-per-minute. The system adjusts for variations in temperature that also affect the speed of sound.

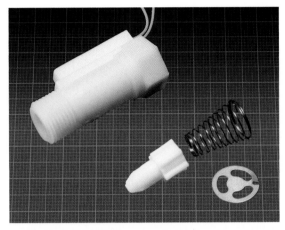

Figure 16-6 *Parts of a flow switch. The small plunger is inserted into the pipe and retained with the perforated plate and compression spring.*

Various configurations are available, some allowing ultrasound sources and detectors to be clamped to the outside of a pipe, as shown in Figure 16-7. To eliminate other variables, one ultrasound pulse is transmitted in the same direction as the flow, followed by another pulse contrary to the flow, and the difference between the two transmission times is used as an indicator of the flow rate.

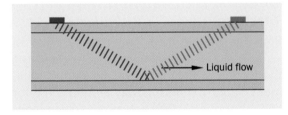

Figure 16-7 *Some ultrasound flow sensors are designed to be clamped externally to a pipe.*

Magnetic Liquid Flow Sensor

A magnetic field is induced in a metal pipe by a coil generating its field perpendicular to flow. The inside of the pipe is lined with nonconductive material in which two electrodes are mounted. Because water containing ions is conductive, the flow of water through the magnetic field induces a small potential difference between the electrodes. This voltage can be used as an indication of the flow rate.

To eliminate external factors that would also affect the potential difference, the polarity of current through the coil around the pipe alternates rapidly. The induced field remains the same regardless of the direction of the current.

A magnetic flow sensor should not be confused with a magnetic flow *switch*. Various types of switches are made, including large, heavy-duty devices where flow moves a magnet that triggers a shutoff valve. This type of industrial device is outside the scope of the Encyclopedia.

Differential Pressure Liquid Flow Meter

In this system, a pipe contains a perforated plate or some similar constrictor that partially obstructs the flow of liquid. Pressure is measured by a pair of pressure transducers placed before and after the constrictor. The pressure difference is an indicator of flow, because it increases as the flow rate increases.

This system was developed originally for large industrial applications but has been miniaturized and etched into silicon to measure very small flow rates. The Omron D6F-PH is an example, measuring less than 3cm square. It contains digital correction to enable a close-to-linear output. Because of its small size, it can be used only for slow flow rates, or as a bypass sensor. The general concept of a bypass sensor is illustrated in Figure 19-7, in the entry discussing gas flow sensors.

What Can Go Wrong

Vulnerability to Dirt and Corrosive Materials

MEMS liquid flow rate sensors containing very delicate, very small sensing elements are very vulnerable to contamination with dirt. Liquids should be filtered to minimize this risk. A manufacturer's datasheet should provide information about the use of corrosive or chemically active liquids.

gas/liquid pressure sensor

17

Pressure measurement devices that do not contain any electronics, such as a nondigital tire-pressure gauge or a mercury manometer, are outside the scope of this Encyclopedia.

Many pressure sensing methods can be used with both gases and liquids. To avoid duplication, gas pressure sensors and liquid pressure sensors do not have separate entries. Both types are described here.

This entry deals almost exclusively with MEMS components. It does not include pressure measurement devices sold as industrial products.

Some manufacturers and vendors use the term *pressure sensor* to describe a component that measures mechanical load or force. In this Encyclopedia, the term only describes components that measure liquid or gas pressure. Mechanical load cells and force sensors will be found in the section on **force** sensors. See Chapter 12.

OTHER RELATED COMPONENTS

- **liquid level** sensor (see Chapter 15)

- **liquid flow rate** sensor (see Chapter 16)

- **gas flow rate** sensor (see Chapter 19)

What It Does

A pressure sensor measures the force exerted by a gas or liquid, often in a container or pipe.

Static pressure is measured under conditions that change slowly or not at all. *Dynamic pressure* is subject to fluctuations. Pressure sensors tend to be designed for one condition or the other.

Schematic Symbols

Many specialized symbols are used in flow diagrams to represent pumps, valves, and sensors, including pressure sensors. Typically they involve a single letter or an X in a circle. These symbols are not generally found in electronic schematics, and therefore they are not included here.

Applications

Barometric sensors are found in barometers and weather stations. Altimeters are really a specialized form of barometric sensor, used in airborne vehicles. Gas pressure sensors have many industrial applications, and are used to monitor tire inflation in vehicles and the output from air compressors. They may also measure liquid pressure indirectly, as in a blood-pressure cuff.

Liquid pressure sensors are widely used to measure oil pressure in automobile engines and hydraulic braking systems. They have medical applications, and monitor water pressure in municipal systems and pumped supplies.

Design Considerations

As a result of random molecular movement, gases will tend to disperse to fill a container. After a gas reaches equilibrium, the pressure will be almost equal in all directions, affected only slightly by gravity. Where a gas is in a sealed, rigid container, pressure will vary linearly with temperature.

A liquid will tend to accumulate at the bottom of any container under the force of gravity. Liquid pressure in a container will be highest at the bottom, because of the weight of liquid above it. However, because almost all liquids are not easily compressible, they transmit force from any point in a container to any other point, including the sides of a container.

Units

Pressure is measured as force per unit area, which can be expressed in a confusing variety of units.

In the United States, gas and liquid pressure are still often expressed in pounds per square inch, abbreviated as PSI, or more often as psi, or sometimes as lb/in^2.

In standard international (SI) units, 1 bar of pressure is approximately equal to atmospheric pressure at sea level. Millibars are popular among meteorologists, 1 millibar being 1/1000 of a bar, equivalent to 100 pascals, where 1 pascal = 1 newton per square meter.

A bar is equivalent to 14.504 psi.

Blood pressure is measured in millimeters of mercury, because mercury manometers were used for this purpose originally. Atmospheric pressure may also be measured in millimeters of mercury, because the earliest barometers used a tube containing mercury. In the United States, some sources still refer to inches of mercury.

How It Works

Pressure sensing usually entails three stages.

1. A *sensing element* translates pressure into the mechanical displacement of a flexible part.

2. A *transducer* converts mechanical displacement into an electrical effect, either modifying resistance or creating a small voltage or current.

3. Electronics are used for *signal conditioning*. This may entail modifying a nonlinear signal, or may convert an analog output to a digital output.

As pressure sensors are increasingly taking the form of MEMS devices, all three stages may be combined in one silicon chip.

Basic Sensing Elements

Figure 17-1 shows four types of sensing elements that are used, or have been used, to convert pressure into mechanical motion. In each case, a green arrow shows where gas or liquid is introduced under pressure. 1: A Bourdon tube flexes under pressure, increasing its radius. The tube is hollow, open at one end and sealed at the other. 2: A coiled Bourdon tube uncoils partially under pressure, causing the top end to rotate. 3: A simple flat diaphragm. 4: A ribbed diaphragm.

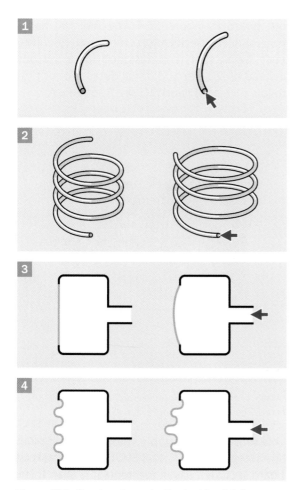

Figure 17-1 *Sensing elements. See text for details.*

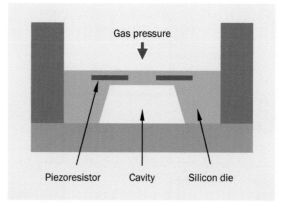

Figure 17-2 *A section of a flat-diaphragm pressure sensor, viewed from the side. An aperture in the top face of the chip allows air to enter. The sensor is etched into silicon, and the deflection of the wafer is measured using embedded piezoresistors.*

Relative Measurement

Pressure is a relative measurement. It is expressed relative to a *reference pressure* of some kind. Three types of measurement are commonly used:

1. *Absolute pressure*, relative to the zero value of a vacuum.

2. *Gauge pressure*, relative to ambient pressure (that is, pressure in the environment around the sensor). Air pressure is the reference source for gauge pressure, and a *vent* is incorporated into the sensing system.

3. *Differential pressure*. In this case, the pressure being measured is relative to some other pressure—for example, the pressure differential between two sealed tanks.

Figure 17-3 illustrates these three measurement types.

The sensing elements numbered 1 and 2 are less likely to be used in modern systems, although a coiled Bourdon tube may be used to turn a potentiometer in low-cost oil-pressure sensing designs. Number 3 is well suited to MEMS devices, as it can be etched into silicon using the principle illustrated in Figure 17-2.

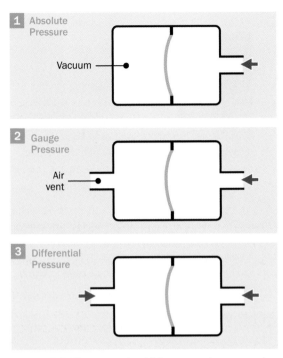

Figure 17-3 *Three ways in which pressure is measured.*

Variants

Ambient Air Pressure

A *barometric* sensor measures the pressure of air around it. This is an absolute value, relative to a vacuum.

The earliest barometers consisted of a tube sealed at one end, containing mercury. No air was allowed to enter the tube. It was inverted so that its open end was immersed in a small reservoir that was open to the air, the pressure of which supported the column of mercury in the tube. Thus, the height of the column was ratiometrically equivalent to atmospheric pressure. Because low air pressure is often an indicator of inclement weather, a barometer was a very simple tool for weather forecasting.

A modern barometric sensor consists of a chip with a vent hole in its upper surface, allowing ambient air pressure to reach a sensor inside the chip. A popular example has been the Bosch BMP085, shown soldered onto a break-

out board in Figure 17-4. The BMP180 has a very similar specification, the primary difference being that it is designed for a SPI bus instead of I2C. For additional details about protocols such as SPI and I2C, see Appendix A.

Figure 17-4 *A Bosch barometric sensor on a breakout board. The background grid is in millimeters.*

This sensor uses a supply voltage ranging from approximately 2VDC to 3.5VDC, but a breakout board includes a voltage regulator that tolerates a 5VDC supply. The output is digitized for access by a microcontroller, but the format is quite complex and consists of raw data that must be converted to air pressure by applying a formula. The manufacturer offers free code in the C language for this purpose. A thermometer that is built into the chip enables temperature compensation.

Subsequent products from Bosch include the BME280, which adds a humidity sensor. Breakout boards using this and other barometric chips are available from sources such as Sparkfun and Adafruit, where Arduino code libraries to interpret the data are available for download.

Altitude

A barometric sensor can be used to determine altitude. At sea level, air pressure is approximately 101kPa (kilopascals), or 760 millimeters of mercury. At 5,000 meters altitude, air pressure drops to 56kPa. The transition is nonlinear, changing most rapidly at lower altitudes.

Air pressure will be affected by temperature and weather conditions. Weather is unlikely to affect the value by more than plus-or-minus 1%, and temperature has a smaller effect; but the barometric sensor can be reset to a zero point based on current temperature and weather data.

Gas Pressure

Thousands of board-mountable gas pressure sensors are available from retail component suppliers. The need to make connections via tubing imposes a minimum size limit on these parts. Many have a through-hole format, but even the surface-mount variants can be relatively large (e.g., 10mm × 10mm). Most tend to have barbed tubing ports, the "barbs" being ridges to retain push-on flexible tubing. Output may be analog or digital, the digital protocol being designed for I2C, SPI, or plain TTL serial buses.

These components are designed for pressures ranging up to 500kPa (kilopascals). Some manufacturers use psi in their specifications, while others use bars and millibars, and a few state gas pressures equivalent to inches of water.

Dual-ported gas sensors for differential pressure measurement can be used for gauge measurement (i.e., relative to ambient air) if one port is left open. Absolute pressure sensors have only one port, as they measure pressure relative to a vacuum maintained inside the chip, behind the diaphragm.

- Some gas pressure sensors can do dual duty as liquid pressure sensors, but others cannot, and datasheets must be consulted carefully, as online vendors

may not include this information prominently.

The sensor in Figure 17-5 by All Sensors Corporation is designed for flexible tubing with 1/16" internal diameter (1.6mm). Its pin spacing is 0.1" (2.54mm), making it breadboardable. This sensor is intended for gases only, but has a "10-inch" water-equivalent rating, meaning that it can tolerate a maximum pressure equivalent to that of a 10-inch (25.4cm) column of water.

When supplied with up to 16VDC, an internal Wheatstone bridge circuit provides an analog output that varies ratiometrically by a few millivolts over the full range of measurable pressures (the exact specification depending on the power supply). The output may be amplified with an external op-amp. Although this is a differential sensor, it can provide gauge pressure if one of the ports is left open.

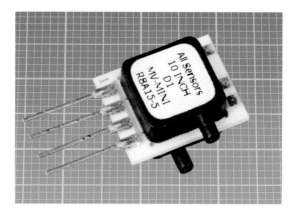

Figure 17-5 *A gas pressure sensor for use with 1/16" (1.6mm) internal diameter flexible tubing. The background grid is in millimeters.*

The larger sensor in Figure 17-6 is in the ADCA range from All Sensors Corporation, designed for flexible tubing with 1/8" internal diameter (3.2mm). This is intended for gases only, and has a "5-inch" water-equivalent rating. It requires a 5VDC power supply and has an internal op-amp that provides an output that varies ratiometrically by 0.2VDC (centered on 4VDC) over the range of measurable differential pressures. Like its smaller cousin, it can provide gauge pressure if one of the ports is left open.

Figure 17-6 *A gas pressure sensor for use with 1/8" (3.2mm) internal diameter flexible tubing. The background grid is in millimeters.*

What Can Go Wrong

Vulnerability to Dirt, Moisture, and Corrosive Materials

MEMS pressure sensors containing very delicate, very small sensing elements must inevitably come into direct contact with gases and, in some cases, liquids. Datasheets will provide warnings regarding humidity and corrosive fluids, but the risk of contamination with dirt remains. Liquids should be filtered to minimize this risk.

Barometric sensors have a hole in the package, exposing the sensor element to the ambient air. This vent must be protected from direct contact with the environment.

Light Sensitivity

If light is allowed into the vent hole in a barometric sensor, it can create photocurrents in the chip. This will affect accuracy.

gas concentration sensor

18

Humidity sensors and *vapor sensors* are included in this entry, as they measure concentrations of a liquid while it is in its gas phase.

Sophisticated *industrial sensors* are available for accurate gas sensing, but this entry deals almost entirely with lower-cost solid-state sensors that are often classified as *board-mount* components.

OTHER RELATED COMPONENTS

- **gas/liquid pressure** sensor (see Chapter 17)

What It Does

Small semiconductor-based gas sensors provide a low-cost method for detecting specific gases in ambient air. The sensors vary their electrical resistance or capacitance in response to the concentration of the gas. They may be used in conjunction with an alarm system that will sound if gases such as *propane* or *carbon monoxide* exceed a preset level, or if the concentration of *oxygen* drops below a preset level.

Because vapor consists of a liquid in its gas phase, gas sensors can respond to vapor, as in the case of *alcohol sensors*.

A *humidity sensor* measures the amount of water vapor in the air, which can be important in refrigeration, HVAC (heating, venting, and air conditioning), medical equipment, meteorology, and storage rooms where art objects, antiques, or paper archives must be preserved in an environment that is neither too dry nor too moist, and is held at a constant temperature. Humidity control can be important also in climates where growth of mold is a concern. Humidity sensors are also used in automobile climate control and windshield defogging, and in the storage of food, fabrics, wood products, and medications.

High humidity coupled with high temperature contributes to the decomposition of many substances, while low humidity can cause desiccation. High humidity can also create damage if it causes materials to expand as a result of taking up moisture.

Many methods have been devised to detect individual gases, but semiconductor sensors are now dominant in applications where accuracy and gas-specificity are not crucial.

Schematic Symbol

No specific schematic symbol exists to represent a semiconductor gas sensor.

Semiconductor Gas Sensors

During the development of transistors in the 1950s, engineers noticed that semiconductor p-n junctions were sensitive to the presence of some gases in the atmosphere. This was regarded as a problem, which was solved by encapsulating transistors to prevent their exposure.

During the 1980s, Japanese law required installation of sensors in every home to detect hazardous concentrations of propane gas. This encouraged the development of cheap, long-lasting components that took advantage of semiconductor sensitivity.

Tin oxide is widely used in a variety of solid-state gas sensors. A sintered layer of the compound is deposited on a ceramic substrate in combination with other compounds such as antimony oxide. The granular layer functions as an n-type semiconductor in which electron transfer increases when certain gases are adsorbed among the grains. When the gas concentration diminishes, oxygen atoms displace the gas molecules, and the sensor returns to its original state. It is unimpaired by being activated, and has a life expectancy of at least 5 years during active (powered) use.

Each sensor includes a tiny resistive heater that is necessary for the chemical reaction to occur. Voltage must be applied to two pins that are connected with the heater. Two other pins connect internally with the sensing element. The resistance between these pins will vary with presence of the target gas; thus, this type of component has a resistive output.

It is a feature of semiconductor gas sensors that they tend to suffer from *cross-sensitivity*. That is, one sensor may respond to more than one gas. Manufacturers control this problem to some extent by adding filtering material around the semiconductor element, or by adjusting the proportions of dopants used in the semiconductor. Datasheets should be consulted carefully to see if a sensor may give false positives as a result of other gases that are likely to be present in the area where it will be used.

Figure 18-1 shows an MQ-5 propane sensor from Hanwei Electronics. The component also responds to methane, hydrogen, alcohol, and carbon monoxide, but with much less sensitivity. The manufacturer recommends calibrating

the sensor by using a resistor ranging from 10K to 47K in series with the output resistance.

Figure 18-1 *A propane gas sensor. The background grid is in millimeters.*

Figure 18-2 shows an MQ-3 alcohol sensor, also manufactured by Hanwei. It has some sensitivity to benzene, but this is unlikely to be a problem, as benzene is seldom present in significant concentrations in ambient air. However, the response of the sensor to alcohol varies with temperature and humidity. Consequently this component can only be used as a "breathalyzer" if the user can be satisfied with an approximate response.

Semiconductor gas sensors are also available for detecting methane, carbon monoxide, hydrogen, ozone, and other gases.

These components are not low-current devices. Typically the internal heater in the Hanwei range has a resistance of slightly more than 30 ohms, and will draw 150mA to 160mA at 5V (i.e., slightly less than 1W). Because the heater is a simple resistive device, it can be used with AC or DC. The output resistance of the sensor can also be assessed with an AC or DC signal.

Figure 18-2 *An alcohol sensor. The background grid is in millimeters.*

A breakout board from Parallax simplifies the use of Hanwei gas sensors. The board is shown in Figure 18-3. It is compatible with carbon monoxide, propane, methane, and alcohol sensors, each of which can be plugged into a socket on the board. Two trimmers establish the sensitivity and the trip point for the sensing element, and the TTL output is then logic-high when a gas is detected, and logic-low otherwise.

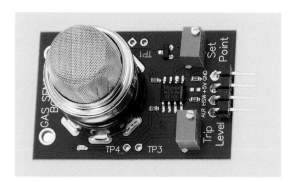

Figure 18-3 *A Parallax breakout board to simplify the use of Hanwei gas sensors.*

Oxygen Sensors

Oxygen sensors are often built using a membrane made of zirconium dioxide. This material has the property that it can transport oxygen

ions when heated. One setup is to have a zirconium membrane separating the gas to be measured from ambient air. This is a type of fuel cell, called a concentration cell or Nernst cell. If the oxygen concentration differs between the two sides of the membrane, oxygen ions will flow through it. Only oxygen ions can move—not neutral oxygen atoms or molecules. The ions are negatively charged, so the transport will lead to a potential difference over the cell, which can be measured with platinum electrodes.

To comply with emissions regulations, automobiles sense the oxygen level of exhaust gases. This data controls the fuel-air ratio in the fuel injection system. Too much air results in the formation of nitrogen oxides, while too little air results in excessive carbon monoxide.

Humidity Sensors

Moisture content of the air is expressed in three different ways:

Absolute humidity
> This is the weight of water vapor in a fixed volume of air. It is measured in grams per cubic meter in the metric system. An absolute humidity sensor is properly called a *hygrometer*.

Dew point
> If a sample of air is cooled without a change of pressure, the dew point is the temperature at which moisture will start to condense. The dew point is a way of describing how humid the ambient air is currently, as water will condense more readily in humid conditions.

Relative humidity
> This is often referred to by its acronym, RH. If temperature, pressure, and volume of a sample of air remain constant, relative humidity is the ratio between the current value of absolute humidity and the hypothetical level where the addition

of water vapor would result in condensation. The ratio is expressed as a percentage. Thus, if moisture is already condensing in an air sample, relative humidity is 100%. If an air sample contains half the weight of moisture required for condensation to begin, relative humidity is 50%. If there is no moisture at all in the air, relative humidity is 0%.

Colloquial usage of the term "humidity" usually means relative humidity, and sensor output is usually convertible to this value. However, some absolute humidity sensors do exist.

Dew-Point Sensor

Historically, meteorologists used a *chilled mirror hygrometer*, in which a metal mirror was exposed to the atmosphere and cooled until the surface was seen to become misty with condensation. The temperature where this occurred was the dew point.

This system is still used in conjunction with an LED and a **phototransistor**. The LED is positioned so that its light reflects from a mirror, directly onto the phototransistor. The mirror is cooled until mist starts to form, which diffuses the reflected light and causes a well-defined change in output from the phototransistor.

Although the formula linking dew point with relative humidity is complex, a simplified approximation is available that is reasonably accurate so long as the relative humidity is 50% or greater. If RH is the relative humidity, t is the current temperature, and t_D is the dew-point temperature at which mist forms:

```
RH = 100 - ( 5 * (t - t_D) ) approx.
```

Although a chilled-mirror dew-point sensor has a reputation for being accurate, it is heavy, expensive, and impractical for most applications outside of meteorology.

Absolute Humidity Sensors

An absolute humidity sensor may use two NTC (negative temperature coefficient) **thermistors** in a Wheatstone bridge circuit. One thermistor is sealed in a compartment containing dry nitrogen, with zero humidity. The other is exposed to the atmosphere. Current passing through the thermistors raises their temperature to at least 200 degrees Celsius. Because heat radiates less efficiently when there is moisture in the air, the exposed thermistor will run hotter, and its resistance will be higher, at higher levels of humidity. This type of sensor is used in clothes dryers and wood kilns, among other applications.

See Chapters 23 and 24 for more information about thermistors.

Relative Humidity Sensors

Two main types of sensing elements are used to measure relative humidity: resistive and capacitive.

In a *resistive* sensing element, a thin layer of polymer, salt, or other hygroscopic substance is deposited on a substrate consisting of ceramic or other unreactive material. When the substance absorbs water, its electrical conductivity increases. Voltage is applied to the sensing element, and AC is used to avoid polarizing it. The current flow is processed externally, with conversion to DC followed by linearization, meaning that the current is processed to establish an almost linear relationship with gas concentration, with temperature compensation factored in. Alternatively, these functions can be performed by hardware built into the sensor, and a digital value for relative humidity can be accessed by an external microcontroller.

In a *capacitive sensor*, once again a thin film of polymer or metal oxide is deposited on a ceramic or glass substrate, but the film functions as a *dielectric* between two metal electrodes that serve as the plates of a capacitor. The dielectric value changes as the film absorbs moisture, and this causes the capacitance to vary, typically by 0.2pF to 0.5pF for each 1% change in relative humidity. This is almost a linear relationship extending over the entire range from 0% to 100% relative humidity.

The actual capacitance value at 50% relative humidity is likely to be between 100pF and 500pF. The sensor may be excited with AC from an external source, or can be incorporated in a chip that derives AC from a DC power supply and provides a digital output.

To determine the dew point or absolute humidity from a value for relative humidity, ambient temperature must also be measured. A chip such as the Si7005 from Silicon Labs includes a temperature sensor with a relative-humidity sensor based around a capacitor in which polyimide film forms the dielectric. If condensation forms, an on-chip heater will cause it to evaporate so that normal operation can resume. Data from the chip is supplied via an I2C interface.

Humidity Sensor Output

When the output is analog, resistance or capacitance of the internal sensing element is available via two pins or solder pads on the sensor. The analog value must be converted to a value for relative humidity by performing a calculation that takes the current temperature into account. The component may or may not include a temperature sensor.

With a digital output, an internal analog-to-digital converter can be accessed by a microcontroller via serial, I2C, or SPI interface. Alternatively, the sensor may communicate values using pulse-width modulation. Either way, the output is a value for relative humidity, calculated on the chip with reference to an onboard temperature sensor.

Analog Humidity Sensor

The Humirel HS1101 is a low-cost analog-output humidity sensor that varies its internal capacitance between approximately 160pF and 200pF as relative humidity increases from 0% to 100%. The response is almost linear, with the

curve steepening slightly when humidity exceeds 80%. The component is shown in Figure 18-4.

Figure 18-4 *A Humirel HS1101 humidity sensor. The background grid is in millimeters.*

The manufacturer claims a recovery time of 10 seconds after 150 hours of condensation. In other words, the performance of the sensor should be restored to its original specification.

A microcontroller can evaluate the output by measuring the charge time of the internal capacitor in the sensor. If the sensing element is wired in parallel with a 10M resistor, this will allow the capacitor to discharge before the microcontroller charges it again. A 220-ohm series resistor should be used between the microcontroller and the sensor, to limit the charge current. This circuit is illustrated in Figure 18-5.

Alternatively, the sensor manufacturer suggests using the capacitance value to control the output frequency of a 555 timer. A counter or microcontroller may be used to count the number of pulses per unit of time.

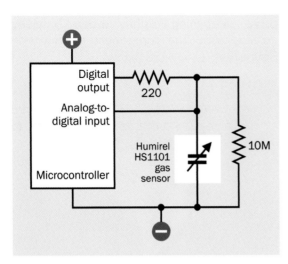

Figure 18-5 *Using a microcontroller to measure the charge time of a gas sensor with capacitive output.*

The Honeywell HIH4030 is a surface-mount humidity sensor with a more convenient analog voltage output that increases almost linearly from approximately 0.8VDC at 0% humidity to 3.8VDC at 100% humidity, assuming a 5VDC power supply. The sensor is available on a miniature breakout board from Sparkfun, shown in Figure 18-6.

Figure 18-6 *The Honeywell HIH4030 on a breakout board from Sparkfun.*

Design Considerations

As relative humidity is temperature-dependent, a relative-humidity sensor must have the same temperature as the air it measures. Many data-

sheets recommend that when a sensor is mounted on a printed circuit board, slots should be milled around it to minimize heat transfer to it. The sensor should also be mounted as far away as possible from heat-generating components.

Where a capacitive humidity sensor with an analog output is located some distance from the electronics that will process its output, shielded cables or twisted pair cables should be used to minimize capacitance in the wiring. A decoupling capacitor between the supply voltage and ground close to the sensor can help to keep the supply voltage stable.

Digital Humidity Sensor

The AM2302, available from Adafruit, is a capacitive humidity sensor with a digital output accessible by a microcontroller via the I2C protocol. Onboard electronics calculate relative humidity with reference to an included temperature sensor. This component is pictured in Figure 18-7.

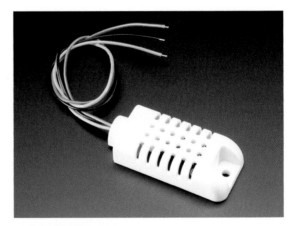

Figure 18-7 *A low-cost humidity sensor with temperature-compensated digital output. From a photograph by Adafruit.*

What Can Go Wrong

Contamination

A semiconductor gas sensor can be damaged by exposure to volatile chemical vapors. This would be unusual in a domestic environment, but is still an important consideration, as there will be no obvious indication that damage has occurred.

Recalibration

If a humidity sensor is exposed to very high humidity where condensation occurs, some datasheets advise a baking procedure, where the sensor is placed in warm, dry air for several hours and then allowed to rehumidify.

Soldering

Semiconductor gas sensors should be soldered quickly and at a controlled temperature, to minimize heat transfer.

gas flow rate sensor

19

Sensors that contain no electronic components are not included in this entry.

A **gas flow rate** sensor may often be described as a *mass flow* sensor or *mass flow rate* sensor. Although it measures volume rather than mass, the mass of gas can be calculated so long as its temperature and pressure are controlled.

Some methods of gas flow sensing can also be applied to liquids, but sensors are usually designed for one application or the other. Therefore, **liquid flow rate** sensors have their own entry. See Chapter 16.

An *anemometer* is a gas flow rate sensor that measures air speed. It is included in this entry.

Many gas flow rate sensors are large devices designed for industrial applications. This entry focuses mostly on lower-cost solid-state sensors that are often classified as *board-mount* components.

OTHER RELATED COMPONENTS

- **liquid flow rate** sensor (see Chapter 16)

What It Does

A **gas flow rate** sensor measures the volume of gas flowing past or through the device, usually inside a pipe. In most applications, users wish to know the mass of the gas that is passing per unit of time. Consequently a gas flow rate sensor is very often identified as a *mass flow rate* sensor. If it functions by heating the gas and measuring the heat dissipation, it is a *thermal mass flow rate sensor*.

A sensor that measures the flow of open air is often described as an *anemometer*. Its output will be expressed as a velocity, not as a volume or mass.

Applications

Mass flow rate sensors are used frequently for laboratory and medical applications, although the reliability and affordability of thermal mass flow rate sensors is now making them attractive for metering municipal gas supplies.

Anemometers are used mostly in meteorology, aviation, and on boats.

Schematic Symbol

Many specialized symbols are used in flow diagrams to represent pumps, valves, and sensors, but they are not schematic symbols of the type generally found in electronic circuit diagrams, and are not included here.

How It Works

Because an anemometer functions so differently from a gas flow rate sensor that is enclosed in a pipe, the two types of sensors will be discussed separately.

Anemometer

The name of this device is derived from the Greek word *anemos*, meaning "wind." The *cup anemometer* was invented in 1846, using four hemispherical cups attached to horizontal arms rotating on a vertical shaft. The rate of rotation was proportional to wind speed over a substantial range, but the conversion factor between wind speed and RPM varied depending on the size of the cups and their distance from the shaft.

Anemometer design was simplified to three cups in 1926 and was modified in 1991 by adding a tag to one cup. This causes the speed of rotation to fluctuate as the tag rotates through the wind stream, and the direction of the wind can be calculated from the speed fluctuations. Not all anemometers rely on this principle, however, and a separate wind vane can be used to determine wind direction.

The basic design of a modern anemometer is illustrated in Figure 19-1.

Figure 19-1 *The basic design of a metereological cup anemometer.*

Anemometers traditionally used a mechanical counter to log the number of rotations, which were checked at fixed intervals to derive the wind speed. The output of a modern anemometer may be achieved by generating AC or DC power, or from *Hall-effect sensors* (see "Hall-Effect Sensor" for a full discussion of Hall-effect sensors).

Handheld Anemometer

A digital handheld anemometer for personal use is shown in Figure 19-2. A cup anemometer made by Vaavud for use with a mobile phone (with appropriate software) is shown in Figure 19-3.

Figure 19-2 *A handheld digital anemometer.*

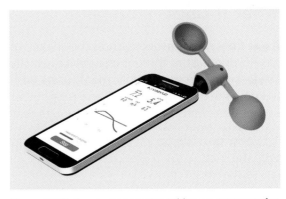

Figure 19-3 *A cup anemometer sold as an accessory for smartphones.*

Ultrasound Anemometer

The movement of air affects the speed of sound, enabling calculation of both wind direction and wind velocity by using an array of ultrasound emitters and receivers. The lack of rotating parts promises greater reliability, and in Figure 19-4 an ultrasound anemometer manufactured by Biral Metereological Sensors also includes heaters so that it will be immune to snow or ice accumulation during freezing conditions.

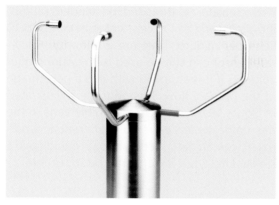

Figure 19-4 *Made by Biral Metereological Sensors, this anemometer uses ultrasound to determine wind speed and direction.*

Ultrasound anemometers have been built on a DIY basis by hobbyists, typically using off-the-shelf ultrasound emitters and an Arduino to decode the signals. Several of these projects are documented online.

Hot Wire Anemometer

A hot wire anemometer measures the cooling effect of the air. It heats a thin wire by passing current through it, and measures the heating power needed to keep the temperature constant.

It is also possible to keep either the voltage over the wire or the current constant, and to assess the wire temperature. The temperature can be measured directly or calculated from the wire's resistance, which increases as the wire gets hotter.

Mass Flow Rate Sensing

A mass flow rate sensor measures the flow speed of a gas. When this is multiplied by the density, the mass flow rate can be calculated.

Most sensors of this type heat the gas and are categorized as *thermal mass flow rate* sensors. The gas passes over a *thermopile* (consisting of several thermocouples wired in series), then a heater, and then another thermopile. These components are miniaturized and can be etched into a chip measuring 2mm square or less.

The temperature difference between the two thermopiles increases as the flow of gas becomes faster and transports more heat to the second thermopile. This is known as the *heat transfer principle*, illustrated in Figure 19-5.

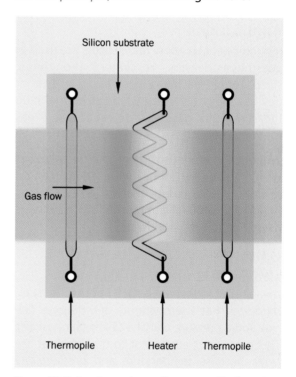

Figure 19-5 *In a thermal mass flow sensor, the flow of gas heats the thermopiles disproportionately, and the temperature difference can be used to evaluate the rate of flow.*

This principle is used also in a liquid thermal mass flow rate sensor, described in "Thermal Mass Liquid Flow Rate Sensor".

An example of a thermal mass flow rate sensor is shown in Figure 19-6.

Figure 19-6 *A thermal mass flow rate sensor made by the Chinese manufacturer Zhengzhou Winsen Electronic Technology Co. Ltd.*

Applications

In medicine, mass flow rate sensors are used in anesthesia delivery, respiratory monitoring, sleep apnea machines, ventilators, and other devices. Industrial uses include air-to-fuel ratio measurement, gas leak detection, and gas metering.

While the metering of municipal gas supplies was traditionally done with all-mechanical devices, MEMS-based meters are replacing many of the 400 million mechanical gas meters estimated to exist worldwide.

Units

Mass flow rate sensors are often rated in SLM, meaning standard liters per minute. A "standard liter" has a temperature of 0 degrees Celsius and a pressure of 101.325 kPa (kilopascals). This pressure is equivalent to that of air at sea level. Because temperature and pressure are specified, the mass of a gas in a standard liter can be calculated by knowing the density of the gas. Thus SLM is a way of measuring mass flow, even though it is a measurement of volume.

The acronym SLPM is also used, but it means the same thing as SLM. SLs and SLPs are measurements of standard liters per second, while SCCM refers to standard cubic centimeters per minute.

Measuring Higher Volumes

MEMS sensors are typically equipped with inflow and outflow nozzles suitable for flexible tubing of 3mm or 5mm internal diameter. The nozzles are "barbed," meaning that they have ridges to retain the tubing.

Small tubing can only deal with low flow rates. A few sensors are threaded for standard plumbing pipes, and can accept volumes of up to 10 liters per minute. They are a minority. A low-rate sensor can still be used to measure higher volumes if it is supplied with just a percentage of the primary flow. This principle is illustrated in Figure 19-7, where an adjustable vane in the main pipe creates a pressure differential. A narrower, constricting section of pipe could have a similar effect, but it would not be adjustable.

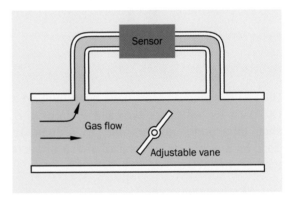

Figure 19-7 *A vane in a primary pipe can divert a percentage of the flow to a sensor.*

Output

Many mass flow sensors have an analog output consisting of a voltage that varies ratiometrically with gas flow. With a typical 5VDC power supply, output voltage may vary from around 1VDC to 4VDC.

Some sensors now incorporate analog-to-digital converters and data processing to provide SLM digital values, accessible from a microcontroller via an I2C interface.

What Can Go Wrong

The primary risk for gas flow rate sensors is damage caused by particles and contaminants in the gas stream. A 5-micron filter is recommended. A *dust segregation system* can also be used, consisting of a small compartment containing semicircular centrifugal chambers. Dust tends to follow the outer edge of the flow path, while the flow sensor is placed on the inner side of the path.

photoresistor

20

A **photoresistor** has a function similar to that of a **phototransistor**, but as its name implies, it is a purely passive component that varies its resistance in response to light.

The term *photocell* was used formerly, but has been displaced by the term **photoresistor**, which describes its function more accurately. The term *photoconductive cell* is sometimes used, or *light-dependent resistor* (or its acronym, *LDR*).

OTHER RELATED COMPONENTS

- **resistor** (see Volume 1)
- **photodiode** (see Chapter 21)
- **phototransistor** (see Chapter 22)

What It Does

Formerly known as a *photocell*, a **photoresistor** is a disc-shaped component with two leads. When light falls on the surface of the disc, resistance between the leads will diminish. Some photoresistors have a resistance in darkness as high as 10 megohms. A few have a resistance in bright light as low as 500 ohms, although several kilohms would be more common.

A photoresistor is less sensitive to light than a **phototransistor** or **photodiode**, and unlike them it is a passive component with no polarity. It presents equal resistance to current in either direction, and may be used with DC or AC.

Because cadmium sulfide is commonly used in this component and is regarded as hazardous to the environment, photoresistors are now unavailable in some regions (notably, Europe). However, at the time of writing, they are still available from many Asian sources, and from some importers in the United States.

Schematic Symbol

Six schematic symbols for a photoresistor are shown in Figure 20-1. They are functionally identical, regardless of whether the single slanting arrow across the resistor symbol is omitted.

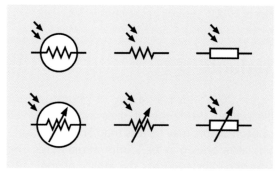

Figure 20-1 *All six symbols for a photoresistor are functionally identical.*

How It Works

Cadmium sulfide is a semiconductor. When exposed to light, more charge carriers are excited into states where they are mobile and can participate in conduction. As a result, electrical resistance decreases.

Construction

A closeup of a photoresistor appears in Figure 20-2. The brown material is a layer of cadmium sulfide deposited onto a ceramic base. The silver material is a conductive compound evaporated onto the cadmium sulfide to form two interlocking electrodes, in a pattern that maximizes the length of the boundary between each of them and the semiconductor. The electrodes connect with leads projecting from the back of the component.

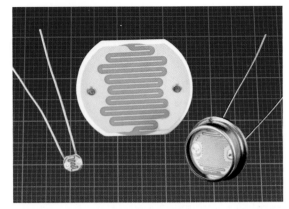

Figure 20-3 *Photoresistors are available in a wide range of sizes. The component in the center is shown at the same scale as the others. Generally speaking, larger components are able to conduct higher currents. The background grid is in millimeters.*

Figure 20-2 *Closeup of a photoresistor. Two interlocking electrodes are mounted on a brown semiconductor layer.*

Variants

Figure 20-3 shows a variety of photoresistors, illustrating the range of sizes available. Small photoresistors may be less than 5mm in diameter; large ones may be 25mm in diameter. The size generally suggests the ability of the component to pass current.

Photoresistors in Optical Isolators

An *optical isolator*, popularly known as an *optocoupler*, contains an LED opposite to a photoresistor, in a sealed package. This is discussed in Volume 2. A *Vactrol* is a similar component, except that the LED is placed opposite to a photoresistor. An example is shown in Figure 20-4.

Figure 20-4 *A Vactrol, containing an LED opposite to a photoresistor. The background grid is in millimeters.*

Vactrol is a brand name owned by its initial manufacturer, Vactec. It was developed to control a vacuum tube, hence its name. In the 1950s, Vactrols were used in guitar amplifiers to control tremolo.

Because of the relatively slow response time of a photoresistor, and its sensitivity to temperature, optical isolators based on photoresistors are not used in digital devices. They still retain some utility in audio components and music equipment, where the ability of the photoresistor to pass AC is an advantage and its response time is adequate.

Values

Datasheets for a few photoresistors can still be found from some suppliers, such as Digi-Key, but are mostly unobtainable, as major semiconductor companies have stopped making these components. Vendors may quote basic values, but in the absence of part numbers or a manufacturer name, a buyer cannot verify the information.

The resistance range can be determined by testing a sample component. A typical range would be from 50K in a light intensity of 10 lux, up to 1M in darkness. Maximum power dissipation is likely to range from 100mW for a small photoresistor to 500mW for a large one.

Maximum voltage may be as high as 200V, but photoresistors will work just as well at low voltages.

Comparisons with a Phototransistor

Slower response

A photoresistor typically takes several milliseconds to respond to bright light, and can require more than 1 second to regain its dark resistance. A phototransistor is much more responsive, and a photodiode is faster still.

Narrower range of resistance

The maximum resistance of a photoresistor is almost always significantly lower than the effective maximum resistance of a phototransistor in darkness, and the minimum resistance is likely to be significantly higher than that of a phototransistor in bright light.

Greater current-carrying capacity

Often a photoresistor is rated for twice as much current as the output from a phototransistor.

Not directional

Because a photoresistor is not packaged with a lens, it is sensitive to incident light from anywhere in front of it. If an application requires that light sensitivity should be confined within a narrow angle while the component is insensitive to incident light from other directions, a phototransistor or photodiode should be used.

Temperature dependent

The resistance of a photoresistor varies more with temperature than the effective resistance of a phototransistor.

Cost

At this time, photoresistors are likely to be more expensive than phototransistors.

Lack of information

Photoresistors are often sold without means of checking their specification in a datasheet.

How to Use It

While the effective resistance of phototransistors and photodiodes varies with the applied voltage, photoresistors present the same resistance for a particular light intensity regardless of the voltage applied. This property has made them suitable for use in "stomp box" guitar-effects pedals.

Because the minimum resistance of a photoresistor in bright light tends to be relatively high, and because its response time is quite slow, it is suitable primarily as an analog component rather than as a switch.

In principle, any resistor in a circuit can be replaced by a photoresistor, and its smooth response to light variations would make it suitable, for example, as the resistance in an RC oscillator circuit, where it would determine the charge time of a capacitor. The frequency of the circuit would thus become light-dependent.

Cadmium sulfide photoresistors respond most actively to wavelengths of light ranging from 400nm to 800nm. This is especially important where an **LED indicator** is used as a light source, as LEDs often emit an extremely narrow range of wavelengths. (See the entry for LED indicators in Volume 2.)

Choosing a Series Resistor

To convert light intensity into a voltage, a photoresistor can be connected in series with a regular resistor, to form a voltage divider. There are two ways to do this, as shown in Figure 20-5. On the left side of this figure, light falling on the photoresistor will cause the output voltage to rise. On the right side, light will cause the output to drop.

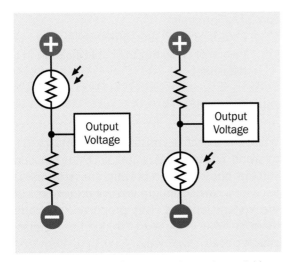

Figure 20-5 *Using a photoresistor to create a variable voltage. See text for details.*

If R_{MIN} is the minimum photoresistor value in the brightest light that will be used, and R_{MAX} is the maximum value in the dimmest light that will be used, a simple formula can be used to find R_S, the optimum value for a series resistor, which will provide the widest range of voltage values at the center of the voltage divider:

$$R_s = \sqrt{R_{min} * R_{max}}$$

What Can Go Wrong

Overload

Because a datasheet may be unavailable, a photoresistor may have to be used on a trial-and-error basis. A test-to-destruction approach may be necessary to determine the limits of the component.

Excessive Voltage

Exceeding the maximum rated voltage for a photoresistor, even for a short time, can cause irreversible damage. Depending on the component, overvoltage can range from 100V to 300V.

Confusion Among Components

Because photoresistors often have no part numbers printed on them and may be sold in miscellaneous assortments, one component may be easily confused with another that has different characteristics. Where two or more photoresistors are used in one device, their characteristics should be measured to determine whether they are functionally identical.

photodiode

21

OTHER RELATED COMPONENTS

- **diode** (see Volume 1)
- **photoresistor** (see Chapter 20)
- **phototransistor** (see Chapter 22)

What It Does

Light falling on a **photodiode** causes it to generate a very small current. This is often called the *photovoltaic effect*. The component functions like a solar panel; in fact, a solar panel can be thought of as being an array of very large photodiodes.

Often, a DC power source is used to apply *reverse bias* to a photodiode. This enables the component to deliver more current. It is now operating in *photoconductive* mode.

Schematic Symbols

Schematic symbols for a photodiode are shown in Figure 21-1. Wavy arrows are often (but not always) used to indicate infrared light.

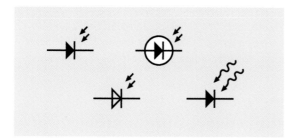

Figure 21-1 *Symbols representing a photodiode.*

Applications

The rapid response of photodiodes makes them suitable for use in optical disc drives, telecommunications, infrared data transfer, digital cameras, and optical switches. Many sensors in this Encyclopedia use a photodiode. **Proximity** sensors, optical encoders, and light meters are examples.

How It Works

When light falls on a semiconductor, it can excite an electron to a higher energy state. The electron then becomes mobile, and leaves behind an *electron hole* (See the entry discussing **diodes** in Volume 1.)

In photovoltaic mode, incident light creates pairs of electrons and electron holes in the semiconductor material. The electrons move to the cathode of the diode while the holes move to the anode, creating voltage between the two. Note that this happens to some extent even in the absence of visible light, as the photodiode may respond to infrared radiation. The tiny amount of current created without visible light is called *dark current*.

In photoconductive mode, light falling on the photodiode creates pairs of electrons and electron holes in the semiconductor material. These will move in opposite directions due to the bias voltage, which means a small current flows through the diode.

Photoconductive mode enables a faster response than photovoltaic mode, because reverse-bias voltage makes the depletion layer wider, which in turn decreases the capacitance. (The same effect is used in *capacitance diodes*.)

Simplified circuits showing the component in photovoltaic mode and photoconductive mode appear in Figure 21-2.

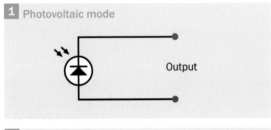

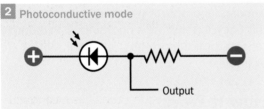

Figure 21-2 *Two modes in which a photodiode may be used. In photovoltaic mode, voltage between the two pins must be measured.*

Variants

PIN Photodiodes

Like *PIN diodes*, PIN photodiodes incorporate an undoped (intrinsic) semiconductor layer between the p- and n-doped layers. They are more sensitive and have faster responses than regular PN photodiodes. Many of the photodiodes available are of the PIN type.

Avalanche Diodes

When light enters the undoped region of the avalanche photodiode, it triggers the creation of electron-hole pairs. When electrons migrate

toward the *avalanche region* of the diode, the cumulative field strength increases their velocity to the point where collisions with the crystal lattice create further electron-hole pairs.

This behavior causes the avalanche diode to be more sensitive than a PIN photodiode. However, the sensitivity also makes it vulnerable to electrical noise, and it is significantly affected by heat. A guard ring is added around the p-n junction, and a heat sink is often used.

Packages

Many surface-mount and through-hole versions are available. A selection is shown in Figure 21-3 and Figure 21-4. While a through-hole photodiode may look indistinguishable from a through-hole 3mm or 5mm LED, versions are available without lenses, and some are sensitive to incident light coming from the side (they are referred to as *side-looking* or *side-view* variants).

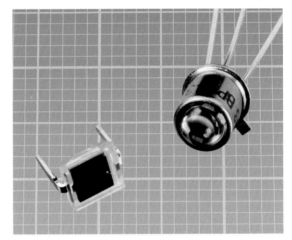

Figure 21-3 *Two sample photodiodes. Left: top-view unfiltered Vishay BPW34. Right: Osram BPX43 in metal can suitable for temperatures up to 125 degrees Celsius.*

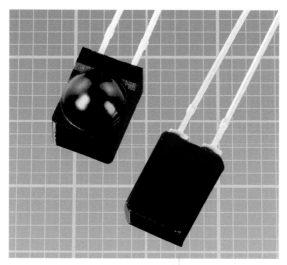

Figure 21-4 *Two side-looking photodiodes. Left: Vishay BPV22NF with lens. Right: Vishay BPW83 without a lens. Both have daylight-blocking filters.*

Wavelength Range

In order for a photon to be detected by the photodiode, it must carry enough energy to be able to create an electron-hole pair. This energy is a property specific to the semiconduct material, and is known as its *band gap*. Additionally, the epoxy packaging of the photodiode may be designed to block some wavelengths of light. Often, an application will require that the component should only respond to infrared light, not visible light.

Photodiode Arrays

A *photodiode array* has several photodiodes mounted in a row or in a grid, for imaging or measurement applications. A row of photodiodes may be used in a flatbed scanner, where it is moved relative to the reflective object being scanned.

In some arrays, photodiodes are available with color filters preinstalled, to facilitate full-color scanning using the transmitted primary colors.

Output Options

Because the output from a photodiode must be processed to be usable, options exist to convert it to a more convenient form, such as a wider range of voltages, a square wave signal where the frequency is proportional with light input, or a binary value accessible by a microcontroller via a serial bus such as I2C. For a discussion of sensor outputs see Appendix A.

Specific Variants

Light to Frequency

The Taos TSL235R is a 3-pin, through-hole chip. It combines a photodiode with logic that creates a square-wave pulse train in which frequency is proportional to the light intensity.

Logarithmic Light Meter

The Sharp GA1A1S202WP light sensor has an output voltage that changes logarithmically with the light level. This gives the sensor a large dynamic range, from 3 lux to 55,000 lux, without requiring a high-resolution analog-to-digital converter. (Human perception of light and sound levels is approximately logarithmic.) This is a surface-mount chip, but is available on a breakout board from Adafruit.

Ultraviolet to Analog

The ML8511 from Lapis Semiconductor combines an ultraviolet-sensitive photodiode with an op-amp that provides an output voltage from approximately 1V to 3V, varying with ultraviolet intensity. A breakout board containing this surface-mount chip is available from Sparkfun and many other sources.

Ultraviolet to Digital

The SI1145 from SiLabs combines ultraviolet sensing with data processing to create a UV index, readable from a microcontroller with I2C. Adafruit offers it on a breakout board.

Color to Digital

The Taos TCS3414FN module contains photodiodes sensitive to red, green, blue, and clear (no filtering). Four 16-bit analog-to-digital converters, one for each channel, provide a digital output accessible over an I2C bus. This module can be used to determine the color of ambient light with some accuracy.

Color to Analog

The Taos TCS3200 also uses red, green, blue, and clear photodiodes but encodes the output from each as a square wave in which the frequency corresponds with the light intensity. The surface-mount chip is available on a breakout board from Robot Shop, shown in Figure 21-5.

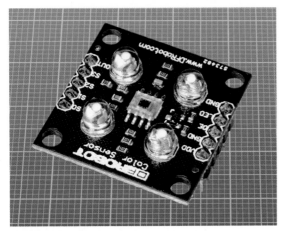

Figure 21-5 *This breakout board from Robot Shop uses the Taos TCS3200 chip to analyze the color of incident light.*

Values

Abbreviations found in datasheets are included in the list below, with values in parentheses for an Osram SFH229FA infrared photodiode, which resembles a 3mm through-hole LED. It has a peak sensitivity of 880nm and appears black to the human eye, being opaque to wavelengths of light shorter than 700nm, the red end of the visible spectrum.

In Figure 21-6 the SFH229FA is shown beside the SFH229, which has the same peak sensitivity of 880nm but is encapsulated in an untinted module, allowing a sensitivity that tapers gradually to below 400nm, in the green part of the

visible spectrum. With the exception of their spectral range, the two photodiodes have identical specifications.

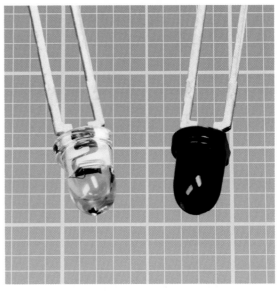

Figure 21-6 *Osram infrared photodiodes SFH229 (left) and SFH229FA (right). The background grid is in millimeters.*

- Typical forward voltage: V_F (1.3V)

- Typical photocurrent: I_P (20µA)

- Maximum power dissipation: P_{TOT} (150mW)

- The *half angle* is measured from the axis of the photodiode to the angle at which the sensitivity has dropped by 50% (plus or minus 17 degrees)

- Dark current: I_R (50pA)

- Wavelength of maximum sensitivity: $\lambda_{S\ MAX}$ (880nm)

- Response speed is the rise and fall time of photocurrent: t_r and t_f (10ns)

Infrared photodiodes exist with a variety of peak wavelengths. They are designed to function in conjunction with an LED that has a matching wavelength.

The angle of sensitivity depends on the geometry of the package.

Rise and fall speeds are important for high-speed measurement, signaling, or data transfer. The rise and fall time of a typical photodiode can be 1,000 times faster than that of a phototransistor. See "Values" in Chapter 22 for a comparison. Note also that the dark current of a photodiode is much lower than that for a phototransistor.

How to Use It

In photoconductive mode, the photodiode can be connected in series with a suitable resistor so that a voltage divider is formed, as shown in Figure 21-2. The voltage at the output will then vary almost linearly with the intensity of the light.

In the photoconductive mode of operation, the output signal is generally measured in millivolts and microamperes. This signal needs to be amplified, usually with an **op-amp** (described in Volume 2).

Figure 21-7 shows simplified schematics for a standard voltage amplifier, in section 1, and a transimpedance amplifier, in section 2.

A transimpedance amplifier measures the current through the photodiode and converts it into a voltage, without the need for a voltage divider. Advantages include less noise, and no need to determine the value of a voltage-divider resistor.

The output voltage of this amplifier is calculated by using this simple formula:

$$V = R * I_P$$

R is the value of the feedback resistor, which determines the gain of the amplifier.

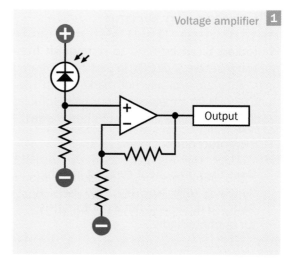

Voltage amplifier 1

Output

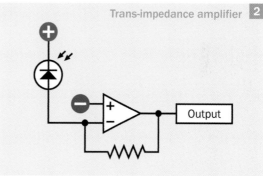

Trans-impedance amplifier 2

Output

Figure 21-7 *Simplified op-amp circuits for use with photodiodes.*

What Can Go Wrong

Photodiodes can be hard to distinguish from phototransistors and LEDs, in particular infrared ones. They are typically not marked with type numbers. Using a meter may not be helpful, as regular LEDs behave similarly to photodiodes.

Decision procedure:

1. Does the component conduct in either direction while shielded from light? If so, it is a diode and not a phototransistor or photodiode.

2. Pass a weak current (e.g., 4mA) in the forward direction. If it emits visible or infrared light, it's an LED. (Infrared light may be visible when using a digital camera, or can be detected with a known phototransistor or photodiode.) If the package is clear but cloudy so that the die cannot be seen, it is probably a white LED in which the cloudiness is the fluorescent pigment converting blue light into white.

phototransistor

22

OTHER RELATED COMPONENTS

- **photodiode** (see Chapter 21)
- **photoresistor** (see Chapter 20)
- **passive infrared** sensor (see Chapter 4)
- **transistor** (see Volume 1)

What It Does

A **phototransistor** is a transistor controlled by exposure to light. (**Transistors** are described in Volume 1.) It can be either a bipolar transistor or field-effect transistor (FET), and its body is often superficially similar in appearance to a 3mm or 5mm **photodiode** or **LED indicator** encased in resin or plastic. (LED indicators are described in Volume 2.) However, some phototransistors are encased in a metal shell with a window in it.

The window or the plastic body is either transparent to visible light, or may appear black if the component is intended for use only with infrared while blocking visible wavelengths. A selection of phototransistors is shown in Figure 22-1 (left: Optek/TT Electronics OP506A with a broad spectral response centered around 850nm; center: Vishay TEKT5400S with a side-view lens; right: Vishay BPW17N).

Typically a phototransistor has two leads that connect internally with its collector and emitter (or source and drain, in the case of an FET). The base of the transistor (or the gate of an FET) responds to light and controls the flow of current between the leads.

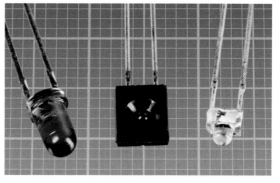

Figure 22-1 *A variety of phototransistors. The background grid is in millimeters. See text for details.*

In the absence of light, a bipolar phototransistor permits leakage between collector and emitter of 100nA or less. When exposed to light, it conducts up to 50mA. This alone differentiates it from a photodiode, which cannot pass much current.

Schematic Symbols

Symbols for a phototransistor are shown in Figure 22-2. They are functionally identical, with the exception of type C, where an additional connection to the base is included. Often (but not always), wavy arrows indicate infrared light.

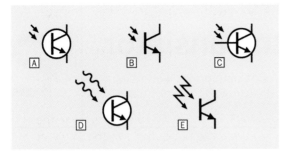

Figure 22-2 *Schematic symbols for a phototransistor.*

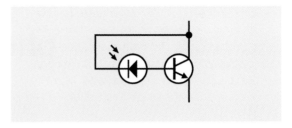

Figure 22-3 *A phototransistor is functionally similar to a photodiode controlling an ordinary transistor.*

Applications

A phototransistor may be used for light measurement or as a light-sensitive switch.

Often the output from a phototransistor is used in conjunction with a *microcontroller* containing an analog-to-digital converter.

Where a clean on-or-off signal is required, a phototransistor can drive the input of a logic chip that contains a *Schmitt trigger*, or it can be processed by a **comparator**.

An *optocoupler* or **solid-state relay** (described in Volume 1) usually contains a phototransistor that is activated by an internal LED. Its purpose is to switch current while electrically isolating one section of a circuit from another.

How It Works

Like a photodiode, the phototransistor responds to light when the light generates electron-hole pairs in the semiconductor material. For a bipolar NPN phototransistor (the most common variant), the important region for pair generation is the reverse-biased collector-base interface. Photocurrent generated here acts as current injected into the base of an ordinary transistor, and permits a larger current to pass from the collector to the emitter.

The behavior of a phototransistor can be visualized as being similar to a photodiode controlling an ordinary bipolar transistor, as shown in Figure 22-3.

Variants

While surface-mount variants are very widely available, through-hole packaging also remains common. When encapsulated like an LED indicator, a phototransistor gathers light from a relatively narrow angle. Variants with a flat surface are sensitive to light from almost any direction in front of the component.

Optional Base Connection

The base of a phototransistor is usually not accessible. However, some variants provide a base connection (or gate connection, in an FET) in addition to the collector and emitter (or the source and drain, in an FET). This third connection allows the application of a bias current, which can prevent low light levels from triggering the transistor.

Photodarlington

A photodarlington is a pair of bipolar transistors, the first being sensitive to light while the second acts as an amplifier for the first. This configuration is very similar to that of a *Darlington transistor* (described in Volume 1). The two-stage design makes it more sensitive to light than a regular phototransistor, but makes the response slower and less linear.

PhotoFET

A field-effect phototransistor is sometimes identified as a *photoFET*. They are relatively uncommon as separate components, but are used in optocouplers, because they have some interesting properties:

- Provided the applied voltage is low enough (less than 0.1V) the photoFET works as a controllable resistance—in contrast to bipolar transistors, where the current is controlled, and is relatively independent of the applied voltage.

- The FET transistor is symmetrical, functioning similarly for signals of either polarity. This makes an FET optocoupler suitable for AC signals.

Values

Abbreviations found in datasheets are included in the list below, with values in parentheses for an Osram SFH300FA infrared photodiode, which resembles a 5mm through-hole LED. It has a peak sensitivity of 880nm and appears black to the human eye, being opaque to wavelengths of light shorter than 700nm, the red end of the visible spectrum.

In Figure 22-4 the SFH300FA is shown beside the SFH300, which has the same peak sensitivity of 880nm but is encapsulated in an untinted module, allowing a sensitivity that tapers gradually to 450nm, in the green part of the visible spectrum. With the exception of their spectral range, the two phototransistors have identical specifications.

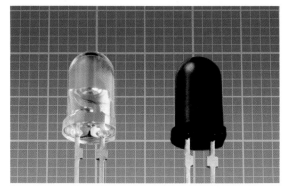

Figure 22-4 *Osram infrared phototransistors SFH300 (left) and SFH300FA (right). The background grid is in millimeters.*

- Maximum collector-emitter voltage: V_{CE} (35V)

- Maximum collector current: I_C (50mA)

- Maximum power dissipation: P_{TOT} (200mW)

- The *half angle* is measured from the axis of the photodiode to the angle at which sensitivity has dropped by 50% (plus or minus 25 degrees)

The angle of sensitivity depends on the geometry of the package. For phototransistors that resemble an LED indicator, with a rounded end that acts as a lens, typical values are plus-or-minus 20 degrees.

- Dark current (when the phototransistor receives no incident light): I_{CE0} (1nA)

- Wavelength of maximum sensitivity: $\lambda_{S\ MAX}$ (880nm)

Infrared phototransistors exist with a variety of peak wavelengths. They are designed to function in conjunction with an LED that has a matching wavelength.

- Response speed is the rise and fall time of photocurrent: t_r and t_f (10 µs)

Behavior Compared to Other Light Sensors

An extended list of comparisons between a photoresistor and a phototransistor will be found in the entry on **photoresistors**. See "Comparisons with a Phototransistor".

A photodiode has a close-to-linear electrical response over a much wider range of intensities of light than a phototransistor. Consequently, photodiodes tend to be the component of choice in applications where measurement of light must be wide-ranging and precise.

Photodiodes can pass less current than a phototransistor, but also tend to draw less current, making them appropriate in battery-powered devices that must draw as little current as possible.

Rise and fall speeds are important for high-speed measurement, signaling, or data transfer. The rise and fall time of a typical phototransistor can be 1,000 times slower than that of a photodiode. See "Values" in Chapter 21 for a comparison. Note also that the dark current of a phototransistor is much higher than that for a photodiode.

The ability of a phototransistor to sink 20mA to 50mA at its output is useful where it will be connected to a component that has relatively low impedance. For instance, a phototransistor can drive a piezoelectric audio transducer directly, or an LED indicator.

Unlike a photodiode, a phototransistor is a solid-state switch. Its saturation voltage (listed in datasheets as $V_{SE(SAT)}$ as described above) is the voltage drop between collector and emitter, and seldom exceeds 0.5V.

Binning

Small variations that occur during the fabrication process can cause inconsistency in the performance of phototransistors that share the same part number. To provide a more consistent response, manufacturers use *binning*, meaning that units sharing the same bin number are likely to share a tighter tolerance. (The same concept is used to minimize variations in **LED area lighting**, described in Volume 1.)

Datasheets will provide information on the availability and meaning of bin numbers, if available.

Bins with higher photocurrents typically have longer response times.

How to Use It

Most phototransistors are bipolar devices with an *open collector output*. That is, the collector of the transistor is accessible from one of the two leads, making it "open" to being used. See Figure A-4 for general instructions on using an open-collector output. A summary relating to phototransistors is included here.

The schematic in Figure 22-5 shows the basic concept. The resistor is referred to as a *pullup resistor*. When the phototransistor is receiving very little light, its effective resistance is high. Consequently almost all the current flowing through the pullup resistor will go to any device attached to the output, and the output voltage will seem to be "high."

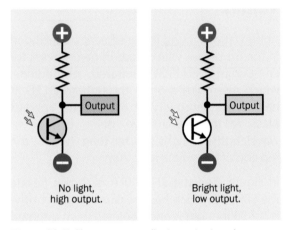

Figure 22-5 *How an open-collector output works.*

If the phototransistor is exposed to a significant light source, the effective resistance between the collector and emitter drops dramatically, and the phototransistor will sink current to ground. Consequently, the output will seem to be "low."

The pullup resistor is necessary between the power source and the collector pin to protect the phototransistor from sinking excessive current when it is exposed to light. The ideal value of the resistor will depend partly on the impedance of any device attached to the output.

In this scenario, exposure to light causes low output whereas darkness causes high output. What if we wish to have it the other way around?

The protective resistor can be moved to the emitter pin, where it becomes a pulldown resistor. It will still protect the phototransistor from passing excessive current, so long as the output is connected to a high impedance. The output is taken from the emitter pin, and will transition from low to high when the component is exposed to light. This is illustrated in Figure 22-6.

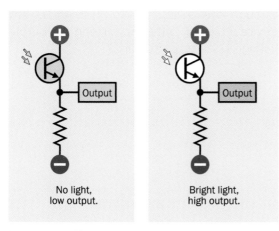

No light,   Bright light,
low output.   high output.

Figure 22-6 *Moving the resistor and taking an output from the emitter pin inverts the behavior of a phototransistor. Any device attached to the output must have a relatively high impedance to protect the phototransistor from excessive current.*

Output Calculation

Using an open-collector output, the photocurrent is almost independent of the applied voltage V_{CE}, provided that the voltage is higher than the saturation voltage $V_{CE(SAT)}$, which is typically between 0.4V and 0.5V.

If the pullup resistor has value R, the voltage across it is

$$U = R * I_P$$

where I_P is the photocurrent passed by the phototransistor.

When choosing R, one should consider the range of currents expected in the light conditions that will be measured, and the voltage range suitable for the next stage of the circuit. 10K is a reasonable starting point (e.g., when measuring light intensity with a microcontroller's analog input). The resistor value can be reduced from there, if necessary.

The value of the pullup resistor must also be chosen to restrict current within the limits specified by the phototransistor datasheet. A value for the resistor guaranteed to be safe is

$$R = V / I_{MAX}$$

where V is the supply voltage and I_{MAX} is the maximum allowed current. With this value, the resistor limits the current to the highest allowed value, even if the phototransistor is brightly illuminated and assumed to conduct perfectly.

When V = 5V and I_{MAX} = 15mA, R should be at least 330 ohms.

What Can Go Wrong

Visual Classification Errors

Phototransistors can be visually similar to LEDs and photodiodes. They are easily confused, as neither type of component is usually marked with any identification number. The entry for **photodiodes** describes a system for distinguishing the three types of components. See "What Can Go Wrong" in Chapter 21.

Output Out of Range

The output voltage from a phototransistor will depend on the intensity of incident light, the value of any pullup resistor that is used, and the supply voltage. While a circuit is being developed, the light range may seem predictable, but in actual use the output range may not fall within expectations.

NTC thermistor

23

PTC thermistors, in which the resistance increases as the temperature increases, have a separate entry. See Chapter 24.

A *resistance temperature detector* or **RTD** has a resistance that increases as its temperature increases, but it is not usually classified as a thermistor, because its sensing element is fabricated differently. Its entry will be found at Chapter 26.

Semiconductor temperature sensors and **thermocouples** each have their own entries.

Infrared temperature sensors and **passive infrared** motion sensors have their own entries. They are *noncontact* temperature sensors that respond to infrared radiation.

OTHER RELATED COMPONENTS

- **PTC thermistor** (see Chapter 24)
- **infrared temperature** sensor (see Chapter 28)
- **passive infrared** motion sensor (see Chapter 4)
- **semiconductor** temperature sensor (see Chapter 27)
- **thermocouple** (see Chapter 25)
- **RTD** (resistance temperature detector) (see Chapter 26)

What It Does

An NTC thermistor is the most common type of discrete-component temperature sensor, and is usually the most affordable. Its resistance diminishes as its temperature increases. This behavior is referred to as a *negative temperature coefficient*, which is the source of the acronym NTC.

This is a simple, passive component that is not polarized. It requires no separate power supply, but an external device must pass a small AC or DC current through it to determine its resistance. This is known as an *excitation current*.

Schematic Symbols

Schematic symbols for a thermistor are shown in Figure 23-1. Those in the top row may still be found in the United States, but are being replaced by the European variants in the second row. The addition of -t° to the symbol indicates an NTC type of thermistor, while +t° indicates that it is the PTC type, with a positive coefficient (see Chapter 24). If no indication is shown, the thermistor is likely to be the NTC type.

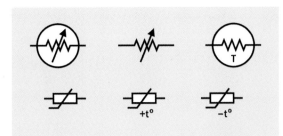

Figure 23-1 *Schematic symbols representing a thermistor. Letter t preceded by a plus or minus sign indicates whether the thermistor is the PTC or NTC type, respectively.*

Applications

Thermistors monitor temperature in air-conditioning systems, clothes washers, refrigerators, pool and spa controls, dishwashers, toasters, and other domestic devices. They are used in laser printers, 3D printers, industrial process controls, and medical equipment.

As many as 20 thermistors may be found in a modern automobile, measuring temperature in locations ranging from the transmission to the ambient air in the passenger compartment.

Comparison of Temperature Sensors

In this Encyclopedia, contact temperature sensors, which measure temperature by making contact with the source, are divided into five main categories, each of which has a separate entry. For convenience, these categories are listed in a comparative summary at the end of this entry. See "Addendum: Comparison of Temperature Sensors".

How an NTC Thermistor Works

Although the term *thermistor* suggests that it is a thermally sensitive resistor, in fact an NTC thermistor is a semiconductor.

Some metal oxides, such as ferric oxide or nickel oxide, become n-type semiconductors

when they are treated with dopants. The exact mix is a proprietary secret of each manufacturer. Raising the temperature of this kind of material increases the number of charge carriers in it, promoting electron mobility and thus lowering its effective resistance.

To create a thermistor, the metal oxide mix is heated until it melts and turns into a ceramic. Typically a thin sheet is cut into small pieces for individual sensors. After two leads are connected, the assembly is dipped into epoxy or encapsulated in glass. The most common packages consist of a glass bead, surface-mount chip, or ceramic disc.

Figure 23-2 shows three NTC thermistors. At left is a Murata NXFT15XH103FA2B100 approximately 1mm in diameter, with a reference resistance of 10K and a tolerance of plus-or-minus 1%. At center is a Vishay NTCALUG03A103GC rated 10K at 2%, fitted with a mini-lug. At right is a TDK B57164K153K rated 15K and 3%.

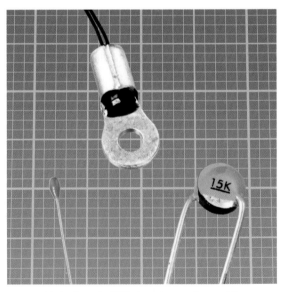

Figure 23-2 *Sample NTC thermistors. See text for details. The background grid is in millimeters.*

Output Conversion for Temperature Sensing

Ideally, the electrical behavior of a temperature sensor should be a linear function of temperature. Thermistors fail in this respect, as their resistance is an approximately inverse exponential function. This is illustrated in Figure 23-3, where the measured resistance of a thermistor rated for 5K at 25 degrees is plotted against temperatures from 0 degrees to 120 degrees Celsius.

- In many datasheets, graphs of this kind may appear flatter because they are customarily plotted against a vertical logarithmic scale.

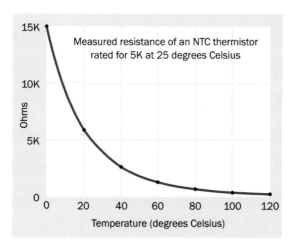

Figure 23-3　*Resistance of a thermistor from 0 to 120 degrees Celsius.*

To monitor the resistance of a thermistor, it can be placed in a simple voltage divider as shown in Figure 23-4, where the fluctuating resistance of the component creates a fluctuating voltage at point A.

- The voltage can be used as an input to a microcontroller that contains an analog-to-digital converter. Alternatively it can be connected directly to a solid-state relay, or amplified with an op-amp, or can be passed through a comparator to create an adjustable switching threshold.

Although this circuit is a voltage divider, it is also known as a *half bridge*, as it is half of a Wheatstone bridge.

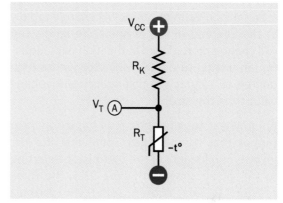

Figure 23-4　*A half-bridge circuit for determining the resistance of a thermistor.*

If V_{CC} is the supply voltage, V_T is the measured voltage at point A, R_T is the resistance of the thermistor, and R_K is the constant value of the series resistor, the basic formula for a voltage divider looks like this:

$$V_T = V_{CC} * (R_T / (R_T + R_K))$$

By transposing terms, a formula can be derived to obtain a value for R_T from the measured voltage and the value of R_K:

$$R_T = (R_K * V_T) / (V_{CC} - V_T)$$

Choosing a Series Resistor

The value for R_K in the formula should be chosen to provide a reasonably wide response over the range of temperatures for which the thermistor is likely to be used. To calculate R_K, another formula must be applied. If R_{MIN} is the resistance of the thermistor at the lowest likely temperature, and R_{MAX} is its resistance at the highest likely temperature:

$$R_k = \sqrt{R_{min} * R_{max}}$$

(This is the same formula as suggested in Figure 20-5 to find the value of a series resistor for use with a photoresistor.)

Wheatstone Bridge Circuit

The half-bridge circuit has the disadvantage that it does not compensate for nonlinearity of a thermistor. Voltage values will change rapidly at the low end of the temperature range, but will change more slowly at the high end, requiring an analog-to-digital converter with a high degree of accuracy to distinguish one voltage value from the next.

A full Wheatstone bridge circuit has a nonlinear output that compensates, somewhat, for the inverse nonlinearity of the thermistor. Referring to the circuit shown in Figure 23-5, the three resistors R_K are chosen using the formula above.

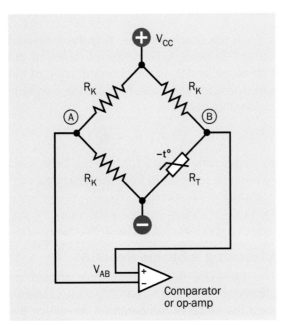

Figure 23-5 *A thermistor may be placed in a full Wheatstone bridge circuit. Outputs A and B are often connected with the two inputs of an op-amp or comparator.*

A standard formula provides the relationship between R_T, the resistance of the thermocouple; V_{CC}, the supply voltage; R_K, the fixed resis-

tances; and V_{AB}, the output voltage measured between points A and B:

$$V_{AB} = (V_{CC} / 2) * (R_T - R_K) / (R_T + R_K)$$

From this formula, a version can be derived to calculate R_T by measuring the output voltage, V_A:

$$R_T = R_K * (V_{CC} + (2*V_{AB})) / (V_{CC} - (2*V_{AB}))$$

- The polarity of V_{AB} is reversible, depending on whether R_T is greater or less than R_K. To accommodate this, A and B can be connected to the two inputs of a comparator or op-amp.

Deriving the Temperature Value

After the resistance of the thermistor has been calculated, it can be converted to a temperature value. The datasheet for a thermistor will usually provide a table showing temperature values tabulated against resistance values, so that a lookup table can be created in a microcontroller program.

Alternatively, a datasheet usually includes constants that can be inserted in a resistance-to-temperature conversion equation, but this is nontrivial and requires natural logarithms, which may not be available in a language implemented on a particular microcontroller.

Inrush Current Limiter

NTC thermistors with appropriate characteristics can be used to limit the inrush of current that tends to occur when a circuit is switched on and large capacitors in the power supply charge very quickly.

An *inrush current limiter* is also known as a *surge limiter*, or may be referred to by its acronym, *ICL*. It is an NTC thermistor whose initial resistance diminishes rapidly as its temperature increases.

While NTC thermistors are the type most often used for inrush limiting, PTC thermistors can serve this purpose if wired differently. See "PTC Inrush Current Limiting". The remainder of this discussion refers only to NTC current limiters.

A suitable NTC thermistor can be placed as shown in the simplified schematic in Figure 23-6, where a rectified AC source is connected with a DC-to-DC converter, and a large smoothing capacitor is used. Initially the thermistor has resistance that is sufficient to limit current and generate heat. But the rise in temperature causes the resistance of the thermistor to fall. Eventually it reaches a steady state where it remains sufficiently warm to maintain a low resistance that imposes a negligible load on the circuit.

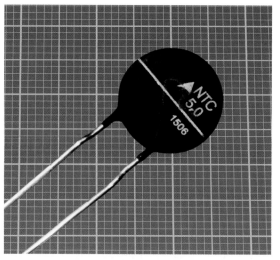

Figure 23-7 *A TDK B57237S509M NTC thermistor designed as an inrush current limiter, rated 5 ohms at 5 amps. The background grid is in millimeters.*

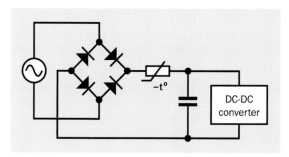

Figure 23-6 *Placement of an NTC thermistor that is designed for inrush current limiting.*

In thermistors that are used for temperature measurement, self-heating is an undesirable attribute. By contrast, an inrush current limiter depends on self-heating to perform its function.

The TDK B57237S509M inrush limiter, shown in Figure 23-7, is rated for 5A and has an initial resistance of 5 ohms at 25 degrees Celsius while not passing current. When tested with a 2,800µF capacitor at 110VAC, its resistance drops to a minimum of 0.125 ohms at 5A. The relationship between current and resistance is shown in Figure 23-8.

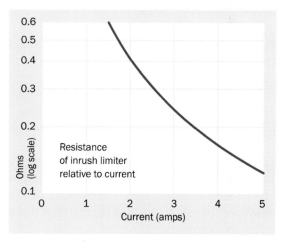

Figure 23-8 *This graph shows the relationship between resistance and current in a TDK B57237S509M inrush limiter.*

Restart

If a protected device is switched off momentarily and is then switched on again, the thermistor cannot provide protection, as it has not had time to cool down and regain its resistance. However, during the 30 seconds to 2 minutes required for heat in the thermistor to dissipate, smoothing capacitors are unlikely to lose much of their charge. Therefore, if the device is restarted, an inrush of current should not occur.

Thermistor Values

Datasheets for thermistors may be more complex and cryptic than for many components.

When examining a datasheet, first check to see if the thermistor is described as being suitable for temperature measurement or inrush current limiting. A component designed for temperature measurement will not survive inrush current, while one designed for inrush current limiting will have a very low resistance making it unsuitable for temperature measurement.

Time and Temperature

In most datasheets, lowercase letter t is used for values relating to time, whereas an uppercase T is used for values relating to temperature. Unfortunately, T may also be used as an abbreviation for "thermistor."

Resistance and Response

Letter R often means resistance, but may indicate response time, depending on the context in which it is used. For example, R_T is the resistance of a thermistor, and t_R is a response time.

Kilohms and Kelvin

Letter K may be used to represent temperature in degrees Kelvin, 0 degrees on the Celsius scale being approximately 273 degrees Kelvin. However, letter K is also used to represent thousands of ohms, sometimes in the same datasheet. In both instances, K is capitalized.

Reference Temperature

This is the temperature at which many attributes of the component are measured, such as its temperature coefficient and resistance. Usually the reference temperature is 25 degrees Celsius, but in some cases it may be 0 degrees, and other values are occasionally used. The term is abbreviated T_{REF}.

Reference Resistance

The reference resistance for a thermistor (sometimes described as its nominal resistance) may be referred to as R_R, and is the resistance at the reference temperature. It may be referred to as the "R value," but in thumbnail product descriptions it can be cited simply as "Resistance."

In datasheets, R25 or R_{25} is the resistance at 25 degrees Celsius. If this is the reference temperature, R_R and R_{25} will be the same.

Dissipation Constant

DC is the *power dissipation constant*, a ratio normally expressed as milliwatts per degree Celsius (written as mW/°C). This is a measurement of how much thermal power the thermistor can transfer to the environment for a 1 degree increase in temperature.

Temperature Coefficient

TC may be used as an acronym for the *temperature coefficient*, which represents the sensitivity of the thermistor. (Sometimes TCR is used instead of TC, the letter R denoting resistance. The two acronyms both mean the same thing.) The value is the percentage change in resistance for each change in temperature of 1 degree Celsius. Thus, if the resistance of a thermistor drops from 800 ohms to 768 ohms when the temperature increases from 28 to 29 degrees Celsius, TC = -4%. For NTC thermistors, which have a resistance that decreases when temperature increases, the temperature coefficient is negative. However, the minus sign may be omitted.

The coefficient may be expressed in parts per million (abbreviated ppm) instead of as a percentage. To convert parts per million to a percentage, divide by 10,000. Thus, a figure of 50,000ppm is equivalent to 5%.

Thermal Time Constant

Unfortunately TC is also used to represent the thermal *time constant*. If T_D is the temperature difference between the thermistor's initial temperature and a new, higher ambient temperature in which it finds itself, TC is the time it takes for the thermistor to add 63.2% of T_D to its current temperature. TC is expressed in seconds, and is defined without power being applied to the thermistor. A low thermal time constant is characteristic of a physically small thermistor that acquires heat rapidly. (TC is very similar to the concept of a time constant for a capacitor acquiring charge. See the entry on **capacitors** in Volume 1.)

Tolerance

The *tolerance* of a thermistor is a measure of its accuracy, usually at 25 degrees Celsius, unless a range of temperatures is stated. A thermistor rated for 5K, with a tolerance of plus-or-minus 1% at 25 degrees, may be found to have an actual resistance ranging from 4,950 to 5,050 ohms at that temperature. Some thermistors have a tolerance of plus-or-minus 20%. A tolerance better than plus-or-minus 1% is relatively rare.

Temperature Range

The *working temperature range* of any thermistor that uses silicon dioxide is usually between about -50 and +150 degrees Celsius (slightly wider for versions encapsulated in glass, and slightly narrower if accuracy is important).

Switching Current

For a thermistor with a nonlinear response, the switching current is the approximate current that forces a sharp transition in resistance. It is represented by I_S.

Power Limitations

Operating current is the maximum current recommended to avoid self-heating. The *power rating* is the maximum allowed power (usually 100mW to 200mW).

Interchangeability

To measure temperature reliably, two thermistors of the same type, from the same manufacturer, should display the same characteristics. This is known as *interchangeability*. A value of plus-or-minus 0.2 degrees Celsius is common

for a modern thermistor, but is not often mentioned in a datasheet.

What Can Go Wrong

Self-Heating

Self-heating can affect the accuracy of an NTC thermistor that is used for temperature measurement. To get accurate temperature readings, keep the current as small as possible. When the resistance of a thermistor is at the high end of its range, brief pulses of current can be used.

Heat Dissipation

Where a thermistor is used for inrush current limiting, it will create some heat during the whole time that a device is switched on. If insufficient air space is allowed between the thermistor and other components, they may be affected.

Lack of Heat

An NTC thermistor will sometimes fail as an inrush current limiter. In very cold climates, it may never become warm enough for its resistance to drop to an acceptable level. Conversely, in a very hot location (such as close proximity to a hot-water pump) it may not get cool enough to provide adequate initial protection.

Addendum: Comparison of Temperature Sensors

A chart illustrating the five main types of contact sensors, and their variants, is shown in Figure 23-9.

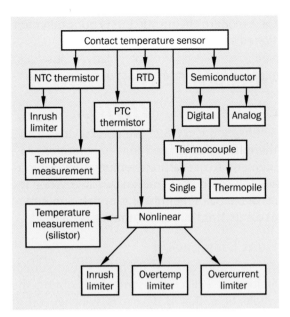

Figure 23-9 *Five types of contact temperature sensors (green boxes) and the variants (red).*

NTC Thermistor

The electrical resistance of an **NTC thermistor** diminishes as its temperature increases. Thus, it has a *negative temperature coefficient*, which is the source of the acronym NTC.

An NTC thermistor is traditionally used where low cost and simplicity are desirable and a relatively limited temperature range is acceptable (often -50 to +150 degrees Celsius). It has the advantage of familiarity, having existed in its present form for many decades. It remains the lowest-cost option among the various types of temperature sensors, and can be connected directly with an external device such as a solid-state relay, in which case no microcontroller is necessary.

PTC Thermistor

The sensing element for a positive-coefficient thermistor is a polycrystalline compound that increases in resistance very rapidly above a threshold temperature. This makes it suitable for blocking a high current to prevent circuit overload.

A *silicon temperature sensor*, sometimes called a *silistor*, can be considered as a PTC thermistor, in that it is a resistive component with a positive temperature coefficient. Its sensing element is etched into silicon.

PTC thermistors are passive, nonpolarized components with two leads or solder pads. For more information about them, see Chapter 24.

Thermocouple

This sensor consists of two wires made from different metals, joined at one end. The differing thermoelectric properties of the wires creates a very small voltage between their free ends. Thermocouples have the widest range of any contact sensor. They are simple, robust, and free from self-heating effects, as they consume no power. Their response is rapid, but very nonlinear, and their sensitivity is limited. They are used in industry and in laboratories, often plugged into a panel meter that combines a digital temperature display with hardware to decode the signal from the type of thermocouple being used.

For more information about thermocouples, see Chapter 25.

Resistance Temperature Detector

Often known by its acronym **RTD**, and sometimes referred to as a *Resistive Temperature Device*, it commonly uses a sensing element fabricated from pure platinum, nickel, or copper. The element may consist of wire wound around a core, or a very thin film deposited on an insulating substrate.

An RTD has a positive temperature coefficient, as its resistance increases while its temperature increases. It is very accurate and stable, providing an almost linear output, especially near the center of its range. However, its sensitivity is often one-tenth of that of an NTC thermistor.

Like a thermistor or a thermocouple, an RTD is a passive device, able to operate at a wide range of voltages and requiring no power supply. It is nonpolarized, with two leads or solder pads.

For more information about resistance temperature detectors, see Chapter 26.

Semiconductor Temperature Sensor

This is a chip-based sensor that requires no additional components to linearize its output, as linearization is performed in the chip.

The temperature range is similar to that of an NTC thermistor, but the output is a variable voltage with a positive temperature coefficient of about 20mV per degree Celsius, supplied by a built-in op-amp. Response time is 4 to 60 seconds.

This type of sensor requires a power supply, typically of 5VDC or less. It does not have to be calibrated before use, as it is trimmed during the production process to achieve accuracy that can be better than that of a thermistor. Manufacturers may claim plus-or-minus 0.15 degrees over the whole temperature range, which is usually -50 degrees Celsius to +150 degrees, but may be less for variants in which accuracy is more important.

The linear analog output is very convenient for use with a microntroller that has an analog-to-digital converter, and the relatively low cost makes this type of sensor increasingly competitive with thermistors.

An analog-to-digital converter may be included on the sensor chip, in which case it is often described as a *digital temperature sensor* or *digital thermometer*, with an output in degrees Celsius (or, sometimes, Fahrenheit) accessible via I2C or SPI bus. For additional details about protocols such as I2C and SPI, see Appendix A.

A *digital thermostat* or *thermostatic switch* is a semiconductor temperature sensor with a binary output that transitions from logic-high to logic-low (or vice-versa) if the temperature goes above a maximum or below a minimum level. The level can be programmed into the chip.

Semiconductor temperature sensors are identified with a variety of other names. For more information, see Chapter 27.

PTC thermistor

24.

A *silistor,* or *silicon-based thermistor,* is included in this entry as a form of **PTC thermistor**.

A *resettable fuse* is not quite the same as a PTC thermistor. For more information, see the entry on **fuses** in Volume 1.

NTC thermistors, in which the resistance decreases as the temperature increases, have a separate entry. See Chapter 23.

A *resistance temperature detector* or **RTD** has a resistance that increases as its temperature increases, but it is not usually classified as a thermistor, because its sensing element is fabricated differently. Its entry will be found at Chapter 26.

Infrared temperature sensors, **semiconductor temperature** sensors, and **thermocouples** each have their own entries.

OTHER RELATED COMPONENTS

- **infrared temperature** sensor (see Chapter 28)
- **semiconductor** temperature sensor (see Chapter 27)
- **thermocouple** (see Chapter 25)
- **NTC thermistor** (see Chapter 23)
- **RTD** (resistance temperature detector) (see Chapter 26)

What It Does

The electrical resistance of a PTC thermistor increases as its temperature increases. Variants can measure temperature or can protect circuits by detecting excessive heat or current.

Because a PTC thermistor is a resistive sensor, it has no polarity. Current may flow through it in either direction, or AC may be used.

Schematic Symbols

The schematic symbol for a PTC thermistor is very similar to the symbol for an NTC thermistor. See Figure 23-1.

Comparison of Temperature Sensors

In this Encyclopedia, contact temperature sensors are divided into five main categories, each of which has a separate entry. For convenience, a comparative summary is included in the entry for

NTC thermistors. See "Addendum: Comparison of Temperature Sensors". Also see Figure 23-9.

PTC Overview

PTC thermistors can be subdivided into two groups:

- *Linear*, with a chip-sized silicon-based sensing element. They are sometimes referred to as *silistors*. The component has a very linear response and is used for temperature measurement. It may be connected directly to a microcontroller.

- *Nonlinear*, mostly using a sensing element containing barium titanate in a polycrystalline compound that increases in resistance very sharply above a threshold temperature. This type of sensor may be described as a *switching thermistor*, because its nonlinear output can activate a switching device.

The sensing elements in positive-coefficient thermistors are different in principle from the element in an NTC thermistor.

Nonlinear thermistors are used in two different ways:

Externally heated
> The thermistor responds to ambient heat or to the temperature of a device to which it is attached. It can be used to protect a circuit or a motor from overheating. Current through the thermistor is minimized to avoid self-heating.

Internally heated
> The thermistor responds to its own temperature caused by current passing through it. It can activate a warning signal or shut down equipment in the event of a short circuit. It can also control current for starting a motor or a fluorescent tube, and is sometimes used as a source to create localized heat.

Silistor for Temperature Measurement

A silicon-based PTC thermistor, sometimes known as a *silistor*, provides a highly desirable, almost linear relationship between temperature and resistance. A popular example is the KTY81 series from NXP, a sample of which is shown in Figure 24-1.

Figure 24-1 *A KTY81 thermistor from NXP. The background grid is in millimeters. Note the amputated center lead.*

The response of this thermistor is shown in Figure 24-2.

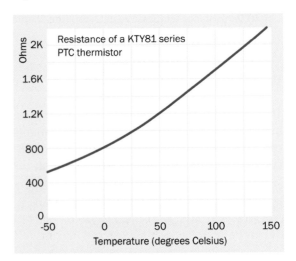

Figure 24-2 *Resistance of the KTY81 thermistor in response to temperature.*

Note that this graph is plotted with a linear vertical scale, unlike the performance curves for many thermistors that are plotted with a log scale. The log scale tends to make a response curve look flatter.

The sensor is a silicon chip designed on the "spreading resistance principle," in which current fans out from a metal contact through a thin layer of silicon to a metallized bottom plane. This effect progresses less actively as the temperature increases. Although the result is partly dependent on polarity, a second metal contact is biased in the opposite direction, and when the two active regions of the chip are wired in series, the result is a component that has no polarity.

- The almost-linear output of this type of sensor makes it easy to use with a microcontroller that has a built in analog-to-digital converter.

Tolerance ranges from plus-or-minus 1% to 5%, depending on the temperature. Variants have a typical reference resistance of 1K or 2K. The temperature coefficient is commonly about 1%, which is considerably lower than that of a typical NTC thermistor, where 4% is common.

- Guidance on reading thermistor datasheets will be found in the entry describing NTC thermistors. See "Thermistor Values".

For correct operation, a typical silistor requires a current ranging from around 0.1mA to 1mA.

- The lower sensitivity and slightly higher price of a PTC temperature-measurement thermistor, compared with an NTC thermistor, may explain why the NTC type seems to remain more popular, with many more variants available. In addition, the NTC type is much more tolerant of variations in current.

- Silistors continue to find some automotive applications, measuring oil temperature, transmission temperature, and climate control, among other parameters.

As a simple strategy to determine its resistance, a series resistor can be used with a PTC sensor to create a voltage divider. The circuit is identical to that used for NTC thermistors. See "Output Conversion for Temperature Sensing".

RTDs

A *resistance temperature detector* or **RTD** is sometimes classified as a PTC thermistor. However, it has a different type of pure-metal sensing element, much lower sensitivity, and is discussed in a separate section of this Encyclopedia. See Chapter 26.

Nonlinear PTC Thermistors

Over-Temperature Protection

This type of nonlinear thermistor is externally heated, but has a switching function. If it is incorporated among other components on a circuit board, its output can be used to activate a warning signal, or can trigger a relay to shut down the circuit until the temperature subsides. This is of special interest for battery chargers where excessive heat can often be a problem, but is also useful in electronic devices generally.

To avoid the possibility of self-heating, current passing through the thermistor must be minimized to a few milliamps.

Some thermistors in the Vishay PTCSL series will make a transition at a temperature as low as 70 degrees Celsius. Others will be triggered by temperatures above 100 degrees. A typical response curve is shown in Figure 24-3, where resistance rises from 100 ohms at 25 degrees to around 1K at the transitional reference temperature of 90 degrees, and reaches at least 4K at 105 degrees.

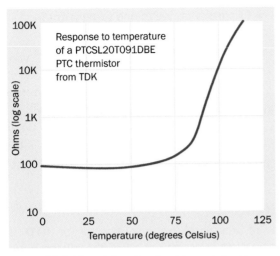

Figure 24-3 *The relationship of resistance to heat in an over-temperature protection thermistor.*

To respond to this transition, the manufacturer recommends a Wheatstone bridge circuit with its outputs connected to a comparator, as suggested for an NTC thermistor in Figure 23-5. The comparator can then activate a signal or a relay.

A picture of the PTCSL20T091DBE thermistor appears in Figure 24-4.

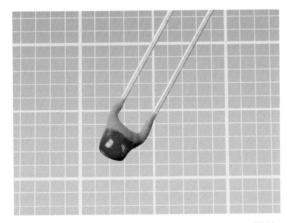

Figure 24-4 *A thermistor in the PTCSL range from TDK. It is color coded using a proprietary scheme by the manufacturer to indicate a reference temperature of 90 degrees Celsius. The background grid is in millimeters.*

This type of thermistor can tolerate a maximum of 30V (AC or DC).

Over-Current Protection

This type of nonlinear thermistor is a substitute for a fuse, as it responds to internal heat created by current passing through it. If the flow of current is excessive, the resistance of the thermistor increases, throttling the flow. When the over-current problem is resolved, the thermistor returns to its normal state. Whereas a fuse must be placed in a location allowing replacement, the thermistor is unharmed by its transition and does not have to be replaced.

Over-current may occur as a result of failure of other components, such as a rectifier diode or a capacitor, or can occur in situations such as a DC motor locking up.

The B598 series from TDK can tolerate voltages over 240V, AC or DC. They typically respond when currents exceed 100mA to 1A, depending on the specific component (a few fall outside that range), and many can withstand 1A to 7A. The B59810C0130A070 pictured in Figure 24-5 is switched by 980mA, can tolerate as much as 7A, and has a reference resistance of 3.5 ohms, rising above 10K when excessive current causes sufficient heat.

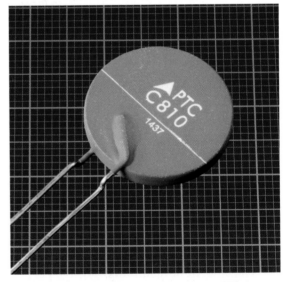

Figure 24-5 *A large over-current protection PTC thermistor. The background grid is in millimeters.*

An over-current thermistor of this type remains wired into the power supply for a device on a permanent basis. Its reference resistance will generate some heat, which is why this type of component is usually restricted to applications where the triggering current is below 1A.

The Murata PTGL07BD220N3B51B0 pictured in Figure 24-6 provides over-current protection with a reference resistance of 22 ohms, has a trip current of 200mA, and tolerates a maximum of 1.5A.

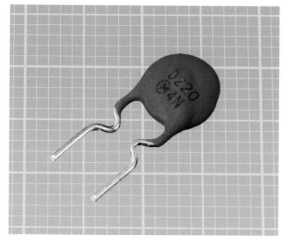

Figure 24-6 *An over-current protection PTC thermistor with a trip current of 200mA. The background grid is in millimeters.*

PTC Inrush Current Limiting

This nonlinear thermistor responds to internal heat caused by an inrush of current when power to a device is switched on. The inrush occurs when current flows rapidly into smoothing capacitors, charging them very rapidly. This can overload a power supply and shorten its life expectancy.

NTC thermistors are traditionally used as inrush limiters. The initially high resistance of this type of component blocks the surge in current, but as heating occurs, the resistance of the NTC thermistor drops rapidly. It remains in the circuit, imposing a relatively small load while the device functions normally. For more details of this application, see "Inrush Current Limiter" in the entry discussing NTC thermistors.

However, an NTC thermistor used in this way will waste some power. Suppose a supply of 120VAC is being used. If the power consumption of a device is 1,000W, the current will be approximately 8A. An NTC thermistor that has a resistance of 0.2 ohms while running warm will introduce a voltage drop of approximately 1.6V, consuming about 13W. This power loss will be greater in applications where the current is even higher—for example, in an electric vehicle recharging station.

To eliminate the loss, a timed bypass relay can be added around the thermistor. The relay closes automatically after a short interval, eliminating the power loss. This is known as *active inrush current limiting*.

However, in this arrangement, an ordinary resistor could be used instead. But in that case, why not use a PTC thermistor that has a reference (cold) resistance of 50 ohms or more? This not only limits the inrush current, but provides additional protection. If a smoothing capacitor in the circuit suffers a short circuit, or if the bypass relay fails to close, excess current passing through the PTC thermistor quickly raises its resistance, protecting the rest of the circuit.

The B5910 series of PTC thermistors from TDK is designed for inrush current limiting. They are packaged in a flame-retardant phenolic resin plastic case, as shown in Figure 24-7. The B59105J0130A020 has a reference resistance of 22 ohms that rises quickly beyond 10K when the temperature exceeds 120 degrees Celsius, as shown in Figure 24-8. This type of component is robust enough to withstand a complete short circuit across a 220-volt supply.

Figure 24-7 *This inrush current-limiting PTC thermistor by TDK is packaged in a flame-retardant case. The background grid is in millimeters.*

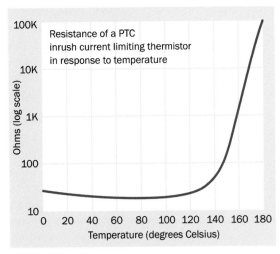

Figure 24-8 *Relationship of resistance to temperature in a PTC inrush current limiting thermistor. Note that the vertical axis has a logarithmic scale.*

PTC Thermistor for Starting Current

In some applications an initial inrush of current is actually necessary and desirable. An air conditioning compressor, for example, requires a surge of high current for "torque assist" when it is starting from a rest state.

High-current PTC thermistors may be used in this kind of situation. The Vishay PTC305C series is an example. These are heavy-duty components that have a switching time of about half a second, a maximum voltage rating of 410VAC or more, and a current rating from 6 to 36 amps.

The PTC thermistor has a relatively high temperature while the motor is running, and must be allowed time to cool before a restart is possible after shutdown. A waiting time of 3 to 5 minutes is imposed by a thermostat or separate time-delay relay.

PTC Thermistor for Lighting Ballast

The starting sequence for a fluorescent lamp requires that current should flow through the cathode heater initially. The thermistor allows this by bypassing a capacitor. Within less than a second, the resistance of the thermistor rises to block current. By this time, the heater has done its job and the lamp runs from high-frequency AC.

PTC Thermistor as a Heating Element

For small applications, a heating element can be made from a PTC thermistor, using its internal resistance to create heat. It has the advantage of being self-limiting, as its resistance rises with temperature. The TDK 5906 series is an example, shown in Figure 24-9. The component is approximately 12mm in diameter, and is designed to be clamped in place, not soldered. It has automotive applications for diesel fuel preheating and spray nozzle defrosting. Residential applications include vaporizers for air fresheners.

The initial resistance is as low as 3 or 4 ohms, rising very quickly at a transition temperature ranging from 70 to 200 degrees Celsius, depending on the specific component.

Figure 24-9 *This TDK B59060A0060A010 heating element is a PTC thermistor whose resistance rises rapidly around 80 degrees Celsius. Rated for 12VDC, it is intended for automotive applications. The background grid is in millimeters.*

What Can Go Wrong

Self-Heating

Self-heating may affect the accuracy of a temperature sensor. To get accurate readings, keep the current small. When the resistance of a thermistor is at the high end of its range, brief pulses of current can be used.

Heating Other Components

In cases where the self-heating of thermistors serves a useful purpose, as in surge protectors and when used for delays, the heat can damage nearby components or materials.

thermocouple

25

Because a *thermopile* is an assembly of thermocouples, it is included at the end of this entry. Other types of temperature sensors have their own entries.

OTHER RELATED COMPONENTS

- **NTC thermistor** (see Chapter 23)

- **PTC thermistor** (see Chapter 24)

- **semiconductor** temperature sensor (see Chapter 27)

- **RTD** (resistance temperature detector) (see Chapter 26)

- **infrared temperature** sensor (see Chapter 28)

What It Does

A **thermocouple** measures temperature by using a pair of wires made from dissimilar metals. At one end of each wire, they are joined together, often by welding them. The differing thermoelectric characteristics of the wires generates a very small voltage between their free ends, from which the temperature of the joined ends can be derived.

No power supply is needed for a thermocouple, but the voltage that it generates is extremely small (measured not just in millivolts, but microvolts) and very nonlinear, requiring hardware and/or software to convert it to a temperature value. Laboratory equipment or integrated circuit chips are available for this purpose.

Different types of thermocouples are available to measure different temperature ranges, and each type has its own characteristics, requiring appropriate conversion.

A "raw" thermocouple looks very unimpressive, as it merely consists of two wires welded together at one end. This is illustrated in Figure 25-1. The full length of the photocouple is shown in Figure 25-2.

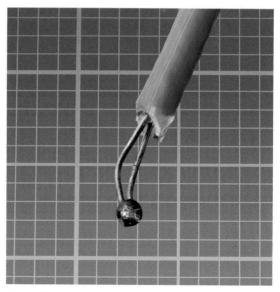

Figure 25-1 *Closeup of the welded wires in a K-type thermocouple. The background grid is in millimeters.*

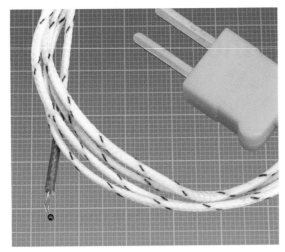

Figure 25-2 *Overview of the thermocouple in the previous photograph.*

A thermocouple sold as a commercial product is likely to be enclosed in a probe, as shown in Figure 25-3.

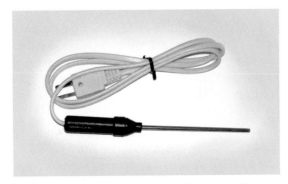

Figure 25-3 *A probe that contains a thermocouple.*

Schematic Symbol

A schematic symbol that is often used to represent a thermocouple is shown in Figure 25-4. Because this component does not consume current, the plus and minus signs do not mean that power should be applied to the wires. The positive sign indicates which wire will generate a higher voltage than the wire with the negative sign.

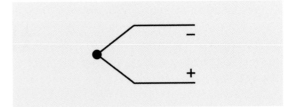

Figure 25-4 *A symbol that is often used for a thermocouple.*

Comparison of Temperature Sensors

In this Encyclopedia, temperature sensors are divided into five main categories, each of which has a separate entry. For convenience, a comparative summary is included in the entry for **NTC thermistors**. See "Addendum: Comparison of Temperature Sensors" for an overview. Also see Figure 23-9.

Thermocouple Applications

Thermocouples have a wider range than any other form of contact temperature sensor, some types being capable of measuring up to 1,800 degrees Celsius. The main limitation is the ability of the joint between the wires to withstand the heat. Appropriate insulation must be used, but segments of ceramic tube are marketed to serve this purpose if necessary.

The very small thermal mass of a thermocouple enables a rapid response to temperature fluctuations. No self-heating occurs, because the thermocouple consumes no power. It is simple and robust. However, its response is very non-linear, and the tiny voltages involved are vulnerable to corruption by electrical noise. Accuracy is usually not better than plus-or-minus 0.5 degrees Celsius, and may be less at low temperatures.

Thermocouples are commonly found in laboratories and in some industrial applications, such as monitoring the temperature in a blast furnace or inside an internal combustion engine.

They may also measure temperatures as low as -200 degrees, but at temperatures below -100 degrees the temperature coefficient diminishes to the point where voltage increments are less than 30µV per degree Celsius.

How a Thermocouple Works

When one end of a piece of wire is maintained at a temperature that is different from the other end, the temperature gradient along the wire creates a small electromotive force that manifests itself as a difference in electrical potential between one end of the wire and the other. This is known as the *Seebeck effect*, named after the man who discovered it. The magnitude of the potential will depend on two factors: the temperature difference between the ends of the wire, and the type of wire that is used.

Figure 25-5 illustrates the concept. Part 1 of this figure shows two wires, named A and B. The left ends of the wires are heated to the same temperature, T_X, while the right ends remain at a cooler temperature, T_Y. Because the wires are composed of different metals, the voltage drop across each wire will be different.

To make this model useful, some factors must be eliminated. In part 2 of Figure 25-5, the hot ends of the wires have been welded together. This now guarantees that they share the same temperature and the same voltage, V_X. We do not yet know what these X values are.

In part 3 of the figure, the cold ends of the wires are clamped in an isothermal block, which keeps them at an equal temperature, still represented as T_Y. The block is not electrically conductive, so the cold ends of the wires still have different voltages, V_A and V_B. We cannot measure these voltages directly, because they are relative to V_X, which is unknown. However, a volt meter can measure V_A and V_B relative to each other.

The volt meter will have its own voltage gradient on its wires, and possibly a temperature

gradient too, but both of these wires are made of the same metal (probably copper) and share the same temperature gradient. Therefore, their effects will be equal.

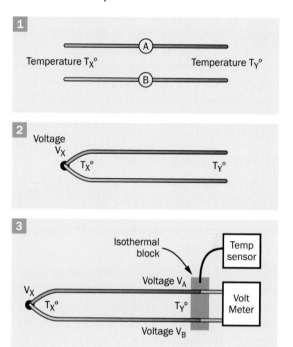

Figure 25-5 *Basic principles of a thermocouple. See text for details.*

A mathematical relationship exists between the temperature gradient and the voltage difference in each thermocouple wire. Suppose K_A is a constant or function that enables the voltage difference in wire A to be determined from its temperature gradient, and K_B serves the same function for wire B. Suppose T_{DIF} is the difference in temperature between T_X and T_Y. We may state:

$$K_A * (T_{DIF}) = V_X - V_A$$

$$K_B * (T_{DIF}) = V_X - V_B$$

By subtracting the second equation from the first and rearranging the terms, we get:

$$T_{DIF} * (K_A - K_B) = V_X - V_A - V_X + V_B$$

The two V_X terms cancel out, leaving $V_B - V_A$ on the right. We can recognize $V_B - V_A$ as the voltage difference measured by the meter. Call it V_M. So:

$$T_{DIF} = V_M / (K_A - K_B)$$

This enables calculation of the temperature difference between the ends of the wires, based on the meter reading and the conversion factor for each wire, which can be found experimentally. Because T_Y is being held at a known, constant value, we can determine the value of T_X:

$$T_X = T_Y + T_{DIF}$$

Thermocouple Details

When the thermocouple was first invented, the cold ends of the wires were placed in a bath of ice and water, forcing them to acquire and maintain a known temperature of 0 degrees Celsius.

The advent of accurately calibrated thermistors made it possible simply to measure the temperature of the cold ends. In this way, a thermistor enables a thermocouple to work. This prompts the question: why not just use the thermistor to measure T_X, and throw away the thermocouple? The reason is that a thermistor has a more limited range, seldom being used for temperatures above 150 degrees Celsius.

Note that the "hot end" of the thermocouple wires does not actually have to be hotter than the "cold end," even though those terms are commonly used. The equation to find T_X works just as well if T_Y is higher than T_X. The temperature difference will simply have a negative value instead of a positive value.

Because "hot" and "cold" are misleading terms, modern documents generally refer to the "measurement junction" and the "reference junction" of the wires. Note, however, that the wires are not actually joined with each other at the reference junction.

A common misconception is that voltage is generated where the wires are joined at the measurement junction. This is not correct. The voltage is a function of the temperature gradient between the measurement junction and the reference junction in each wire. Therefore, the way in which the wires are joined is irrelevant, provided there is an electrical connection between them. They can be welded, soldered, brazed, or crimped together.

How to Use It

Where a thermocouple is used in a laboratory, typically each wire is insulated, and they terminate in a plug that is inserted in a meter. The reference junction is hidden inside the meter, along with some electronics to decode the temperature data. The meter must have a setting that is appropriate to the specific type of thermocouple being used, so that the conversion factors are correct.

Because the type of metal in each wire must be consistent all the way from the measurement junction to the reference junction, other types of wires cannot be used to extend the reach of a thermocouple. Any extension must use wires made from the same metals. Connectors, also, must have pins and sockets that match the types of metals in the wires.

An extension wire for a thermocouple is shown in Figure 25-6.

Figure 25-6 *Extension wire for a type K thermocouple. Note the polarized plug.*

Types of Thermocouples

Thermocouples are identified by ANSI standard codes consisting of single letters of the alphabet, shown below. Temperature ranges are approximate, in Celsius, with minimums rounded up and maximums rounded down to the nearest 50 degrees. Some data sources recommend narrower temperature ranges for practical use.

K type

-250 to +1,350 degrees. Most popular type of thermocouple. Positive wire is a nickel-chromium alloy, negative wire is a nickel-aluminum alloy. Commonly used in 3D printers.

J type

-200 to +1,200 degrees. Positive wire is iron, negative wire is a copper-nickel alloy. The iron wire is magnetic and vulnerable to corrosion. This thermocouple is not recommended for low temperatures, even though it is theoretically capable of measuring them.

T type

-250 to +400 degrees. Recommended for cryogenic applications. Positive wire is copper, negative wire is a copper-nickel alloy.

E type

-250 to +1,000 degrees. Most sensitive type, with the highest temperature coefficient. Positive wire is a nickel-chromium alloy, negative wire is a copper-nickel alloy.

N type

-250 to +1,300 degrees. An alternative to the K type, more stable at higher temperatures. Positive wire is a nickel-chromium-silicon alloy, negative wire is a nickel-silicon-magnesium alloy.

R type

-50 to +1,750 degrees. For high-temperature applications. Positive wire is a platinum-rhodium alloy, negative wire is platinum. Very low temperature coefficient.

S type

-50 to +1,750 degrees. For high-temperature applications. Positive wire is a platinum-rhodium alloy, negative wire is platinum. Very low temperature coefficient.

Seebeck Coefficients

Datasheets for thermocouples list the Seebeck coefficient, which is the temperature coefficient caused by the Seebeck effect, measured in microvolts per degree. In other words, the value provided by a Seebeck coefficient is the number of additional microvolts that a thermocouple will generate for an increase in 1 degree Celsius.

Each type of thermocouple has a different coefficient, and because thermocouples tend to have a very nonlinear response, the coefficients will vary with temperature. Figure 25-7 shows the variations for six types of thermocouples, over a range from -400 to +1,400 degrees Celsius. It is important to understand that the vertical scale shows the coefficient for each type of thermocouple—that is, the *change* in voltage, not the *actual* voltage, at temperature values along the horizontal axis.

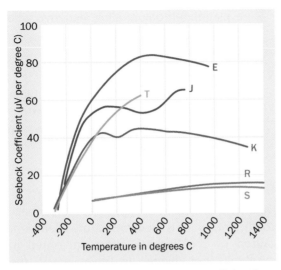

Figure 25-7 *The Seebeck (temperature) coefficient for six types of thermocouples. Partially derived from a datasheet published by Analog Devices.*

R and S type thermocouples have a relatively consistent response, but cannot achieve great accuracy, because the voltage increment is so small for each change in temperature. The K type thermocouple does relatively well between 0 and 1,200 degrees, but the J type only performs adequately between 0 and 800 degrees, while the T and E types are quite inconsistent.

For lower voltage readings, electrical noise becomes an issue. Thermocouple wires are often twisted together, and may also be shielded, to reduce sensitivity to noise. The electronics that decode the thermocouple voltage should include a filter to reject 50Hz or 60Hz interference from wiring in the vicinity.

Chips for Output Conversion

Meters that are designed to interpret the output from a thermocouple and display a temperature tend to be expensive, and may not be convenient for a custom-built application. Fortunately integrated circuit chips are now available to amplify the thermocouple output and apply signal conditioning to create an almost linear response.

The AD8494 and AD8496 from Analog Devices are precalibrated by laser wafer trimming to match the characteristics of J type thermocouples, while the AD8495 and AD8497 match the characteristics of K type. The chips require a power supply of at least 3VDC and have an analog output of 5mV per degree Celsius, enabling measurement over a range of almost 1,000 degrees. They require a very low supply current of 180µA. The manufacturer claims an accuracy of plus-or-minus 2 degrees Celsius.

The chip contains a temperature sensor which should be at the same temperature as the reference junction of the thermocouple. This means that the reference junction (typically, at the socket where a thermocouple plug is inserted) should be as close to the chip as possible, and the chip should be protected from heat created by other components. Any difference in temperature between the reference junction and the chip will create a temperature measurement error.

The MAX31855K chip from Maxim Integrated Products is another thermocouple-to-digital converter. It linearizes the output from the thermocouple and digitizes it as a temperature that is accessible by a microcontroller via a serial SPI bus. Breakout boards with this chip are available. The last letter of the chip number specifies the thermocouple type. Variants for J, K, N, T, S, and R are available.

The AD8495 is mounted on a breakout board from Adafruit, and the MAX31855K is mounted on a breakout board from Sparkfun. These boards are pictured in Figure 25-8.

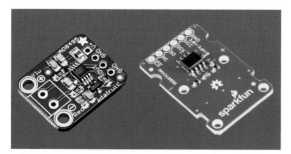

Figure 25-8 *Thermocouple amplifier/conversion chips from Adafruit (left) and Sparkfun (right).*

Thermopile

A thermopile is an assembly of thermocouples connected in series, as suggested in Figure 25-9, where a hot area is shown on the left and a cooler area is shown on the right. The figure assumes that the orange-colored wires have a voltage difference of 5mV from left to right, as a result of the temperature difference, while the purple-colored wires only have a voltage difference of 1mV from left to right. Therefore, the voltage difference between each reference junction and the next is 4mV, as shown on the right side of the figure. So long as the temperature difference exists, the voltage differences will be cumulative, totaling 16mV between top and bottom in this example.

Note that the multiple thermocouple junctions are not electrically connected with each other in each temperature zone.

In reality, more thermocouples than this will be added, and the voltage differences may be lower.

Generally a thermopile is not sold as a separate component from retail suppliers, but is built into other devices. It may be used to generate small amounts of current from a heat difference, as in an infrared thermometer. It can also be used as a safety device to shut down a gas supply if a burner is not lit. See Chapter 28.

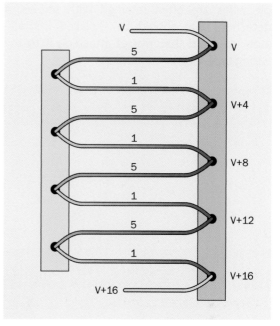

Figure 25-9 *The operating principle of a thermopile. The numbers represent mV, but are chosen arbitrarily as an example.*

What Can Go Wrong

Polarity

The output from a thermocouple has polarity. If this is not observed, an error will result.

Electrical Interference

Thermocouple wires are vulnerable to electrical interference, and should be a twisted pair or, ideally, shielded.

Metal Fatigue and Oxidation

The wire used in some thermocouples tends to be relatively brittle, and cannot withstand much flexing. Also, some metals or alloys are vulnerable to oxidation.

Using the Wrong Type

Different types of thermocouples have totally different characteristics. The electronics to decode the signal from a thermocouple must be matched to the type of thermocouple being used. The plugs on the ends of the thermocouple wires are often retained with screws. A detached plug should be replaced immediately, to avoid the error of attaching it to the wrong type of thermocouple.

Heat Damage from Creating a Thermocouple

If a thermocouple is made from two wires on a DIY basis by welding the tips of the wires together, minimal heat must be used to avoid degrading the alloys in the wires.

RTD (resistance temperature detector)

28

RTD is an acronym, either for *resistance temperature detector* or *resistive temperature device*. No definitive information seems to exist regarding which term is correct, but *resistance temperature detector* is more common.

Occasionally an RTD may be described as a *PTC thermistor*, but its sensing element is different, consisting of pure metal wire or film.

OTHER RELATED COMPONENTS

- **thermocouple** (see Chapter 25)
- **NTC thermistor** (see Chapter 23)
- **PTC thermistor** (see Chapter 24)
- **semiconductor** temperature sensor (see Chapter 27)
- **infrared temperature** sensor (see Chapter 28)

What It Does

A *resistance temperature detector*, also known as a *resistive temperature device*, is usually referred to as an **RTD**. It has a positive temperature coefficient (that is, its resistance increases as its temperature increases) but differs from a **PTC thermistor** in that its sensing element is pure metal instead of a semiconductor.

Comparison of Temperature Sensors

In this Encyclopedia, temperature sensors are divided into five main categories, each of which has a separate entry. For convenience, a comparative summary is included in the entry for **NTC thermistors**. See "Addendum: Comparison of Temperature Sensors" for an overview. Also see Figure 23-9.

RTD Attributes

Positive attributes of RTDs include:

- Accuracy, often plus-or-minus 0.01 degrees Celsius. This very small tolerance allows excellent interchangeability.
- Stability, with a response that drifts by as little as 0.01 degree per year.
- The output is an almost linear function of temperature, making them easily used with a microcontroller.
- Immunity to electrical noise.
- Reasonably rapid response to temperature changes (about 1 to 10 seconds).

Undesirable attributes include:

- Temperature coefficient about one-tenth of that of an NTC thermistor.

- To measure the resistance, some current must pass through the sensor, raising the possibility of self-heating (as in the case of other temperature sensors, with the exception of a thermocouple).

- Relatively high cost, especially of the wire-wound type.

The resistance curves for three generic NTC thermistors are shown in Figure 26-1, plotted against the resistance for a generic platinum RTD that has a reference resistance of 100 ohms at 0 degrees Celsius. Note that this graph is unlike many that illustrate the response of NTC thermistors, in that its vertical scale is not logarithmic.

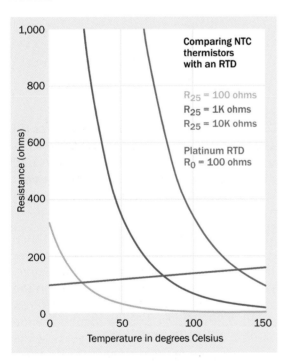

Figure 26-1 *The brown, green, and blue curves show the resistance of three generic NTC thermistors varying with temperature. The red line shows the resistance of a platinum RTD. Derived from a diagram created by Texas Instruments.*

Schematic Symbol

There is no specific schematic symbol for an RTD. Often the symbol for a thermistor may be used. See Figure 23-1.

Applications

Because of its high accuracy, an RTD may be used where precision is important. It can calibrate other temperature sensors, and may measure the temperature of a reference junction of a thermistor. However, it requires sensitive electronics for signal conditioning, because of its low temperature coefficient.

How It Works

An RTD exploits the fractional increase in electrical resistance of a metal film, metal filament, or (in some cases) a carbon film, when the temperature of the metal rises. In its simplest form, an RTD is a 2-wire device with no polarity.

The sensing element is often made from platinum, as this has a linear response to temperature over a wide range. High-quality RTD sensors with a wide temperature range usually consist of platinum wire that is wound around a glass or ceramic core. Smaller sensors may be fabricated from a thin layer of platinum evaporated onto an insulating substrate. Nickel may be substituted for platinum, and has a more sensitive but less linear response.

The wire-wound type is usable at temperatures as high as 500 Celsius (up to 1,000 degrees for some platinum-element types). Some variants are able to measure temperatures as low as -250 degrees.

DIN 60751 is an international standard defining the performance of platinum RTDs. It specifies a reference resistance of 100.00 degrees at 0 degrees, and a temperature coefficient between 0 degrees and 100 degrees. Outside of this range, a formula defines the response.

The response is almost precisely linear, ranging from 100 ohms at 0 degrees to approximately

138 ohms at 100 degrees. The temperature deviation from a straight line, between 0 and 100 degrees, is no more than plus-or-minus 0.8 degrees.

However, current through the RTD must be restricted to avoid self-heating. A range of 0.5mA to 1mA is recommended.

Variants

Some RTDs are potted in glass or resin, as shown in Figure 26-2. This shows an RTD in the TFPTL range from Vishay, containing a nickel thin-film sensing element with a temperature coefficient of about 0.4% and a tolerance of 0.01%. It is available with a very wide range of reference resistances, from 100 ohms to 5K (measured at 25 degrees Celsius). The temperature range is -55 to +70 or -55 to +150 degrees, and the maximum voltage is 30V to 40V, depending on the specific component.

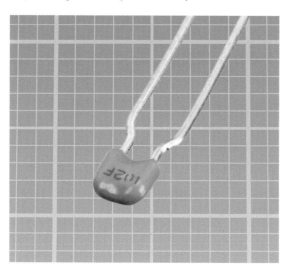

Figure 26-2 *An RTD in the TFPTL series from Vishay. The background grid is in millimeters.*

The flat package shown in Figure 26-3 may be encased in a protective sheath of plastic or silicone rubber, and can be used for surface temperature sensing where the component is glued to the exterior of a flat-sided container. This figure shows an RTD in the L420 range

from Heraeus Sensor Technology, containing a platinum thin-film sensing element with a temperature coefficient of 0.385%. It is available in reference resistances of 100, 500, and 1,000 ohms (measured at 25 degrees Celsius). The temperature range is -50 degrees to +400 degrees.

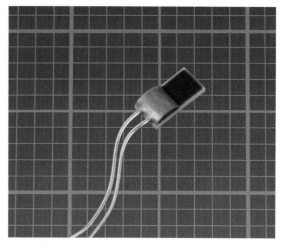

Figure 26-3 *An RTD in the L420 series from Heraeus. The background grid is in millimeters.*

Wiring

The leads to an RTD can be a source of error. If a simple 2-wire configuration is used, the leads will have an unknown resistance that will be affected by temperature, just as the element inside the RTD will have a temperature-sensitive resistance.

To enable temperature compensation, a three-wire design can be used. Figure 26-4 illustrates the principle. In the first section of this figure, resistances R_A and R_B remain unknown. In the second section, the resistances of R_B and R_C can be found by passing a test current through one wire and back through the other, bypassing the component. Assuming that all of the leads are identical in length and composition, the resistance of $R_A + R_B$ will be equal to that of $R_B + R_C$.

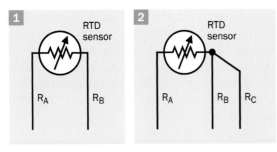

Figure 26-4 *A three-wire configuration enables temperature compensation for the leads to an RTD. See text for details.*

RTD Probe

For practical applications, an RTD sensor is often packaged inside a probe that can be indistinguishable from the type of probe used with a **thermocouple**. However, a thermocouple always uses two wires, as the wires themselves create the voltage. An RTD often uses three wires, as shown in Figure 26-5. This particular sensor is sold for use in a "Brew-Magic" system for brewing craft beer on a commercial basis.

Figure 26-5 *A three-wire RTD is packaged inside this steel probe.*

Signal Conditioning

To process the signal of an RTD, a chip such as the LM75 from National Semiconductor can be used. This is calibrated for connection with a platinum RTD. Internally it converts the resistance of an RTD to a value of 5mV per degree Celsius. This then passes through an analog-to-digital converter on the chip, creating a digital value that can be read via an I2C bus.

What Can Go Wrong

Self-Heating

Self-heating is an issue for RTDs, just as it is for thermistors. Current through an RTD should be limited to 1mA, especially when measuring low temperatures.

Insulation Affected by Heat

The resistance of insulation on the wires leading to a sensor can change with temperature, leading to incorrect resistance readings. This is a more likely source of problems for RTDs than for thermistors, as RTDs are often used at higher temperatures and have a lower temperature coefficient.

Incompatible Sensing Element

If signal conditioning is applied to an RTD that has an incompatible sensing element, temperature readings will be incorrect. For example, an RTD with a nickel element should not be used with signal conditioning designed for a platinum element.

semiconductor temperature sensor

27

This type of sensor may also be referred to as a *bandgap temperature sensor*, a *diode temperature sensor*, a *chip-based temperature sensor*, or an *IC temperature sensor*.

The unfortunate term *integrated silicon-based sensor* is sometimes used, which can be confused with *silicon temperature sensor* (also known as a *silistor*), which is a type of **PTC thermistor**. See Chapter 24.

Some vendors do not divide temperature sensors into clear categories. Semiconductor temperature sensors may be classified as *board-mount* temperature sensors, even though many of them have leads and are not specifically designed to be mounted on circuit boards.

A semiconductor temperature sensor with a digital output is sometimes described as a *digital temperature sensor* or *digital thermometer*. This can be misleading, as the outputs from other types of temperature sensors may be digitized with appropriate components.

OTHER RELATED COMPONENTS

- **thermocouple** (see Chapter 25)
- **NTC thermistor** (see Chapter 23)
- **PTC thermistor** (see Chapter 24)
- **infrared temperature** sensor (see Chapter 28)
- **RTD** (resistance temperature detector) (see Chapter 26)

What It Does

A **semiconductor** temperature sensor is an integrated circuit chip incorporating a sensing element composed of transistor junctions. It has an approximately linear response and is easy to use, in some cases being designed for direct connection with a microcontroller, requiring no additional components.

In analog variants, the output consists either of voltage or current that varies with temperature. These components have a positive temperature coefficient, except for a few CMOS variants where a voltage output diminishes as temperature increases.

Digital variants are becoming more common, creating a numeric output accessible by a microcontroller.

In almost all semiconductor temperature sensors, the characteristics of silicon dioxide limit the temperature range to approximately -50 to +150 degrees Celsius (sometimes less).

This type of component is not (yet) as low in cost as a thermistor, but can include its own amplification, signal processing, and (optionally) analog-to-digital conversion on one chip.

Comparison of Temperature Sensors

In this Encyclopedia, temperature sensors are divided into five main categories, each of which has a separate entry. For convenience, a comparative summary is included in the entry for **NTC thermistors**. See "Addendum: Comparison of Temperature Sensors" for an overview. Also see Figure 23-9.

Semiconductor Temperature Sensor Applications

When a semiconductor temperature sensor is used in surface-mount format, it can measure the temperature of the board on which it is mounted. This enables protection from overheating, often in power supplies.

Because the sensing elements and signal processing circuits are all chip-based, they can be transplanted into other types of sensors. For example, a gas pressure sensor or a proximity sensor can have onboard compensation using a semiconductor temperature sensor. They have also been built into computer CPUs such as the Pentium series from Intel.

Some variants are manufactured in a three-lead TO-92 package, appearing superficially similar to bipolar transistors. They are suitable for remote temperature sensing, and have automotive applications such as measuring the temperature of the transmission, engine oil, or cabin interior. They may also be found in some heating and air-conditioning systems, and some kitchen equipment.

Schematic Symbol

No unique schematic symbol has been developed for a semiconductor temperature sensor. It may be represented by a rectangle contain-

ing text abbreviations to represent pin functions, similar to other types of integrated circuit chips.

In the case of a sensor with an output consisting of current that varies with temperature, the sensor may be shown as a *current source*, using the symbol in Figure 27-1. However, this symbol is not specific to temperature sensors; it is used for any component that is a current source.

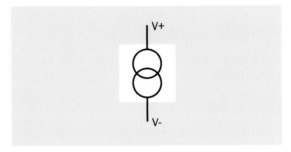

Figure 27-1 *A temperature sensor whose output consists of current varying with temperature may be shown in a schematic as a current source, using this symbol.*

Attributes

Desirable attributes of a semiconductor temperature sensor include:

- Easy to use. Few or no external components may be required, and little or no signal processing.

- Factory-calibrated, with an almost linear response.

- Versions with a digital output are easy to add to any system that already has an I2C bus. For additional details about protocols such as I2C, see Appendix A.

Undesirable attributes of semiconductor temperature sensors include:

- Limited temperature range, the same as thermistors.

- Self-heating issues, especially in versions where signal-processing functions are built into the same chip.

- Not as rugged as some types of temperature sensors.

For an explanation of terminology used for temperature sensors in datasheets, see "Thermistor Values".

How It Works

When a constant current is flowing through a p-n junction in a diode, the voltage across the diode will change by about 2mV for each change in temperature of 1 degree Celsius. This can be demonstrated by the simple circuit shown in section 1 of Figure 27-2.

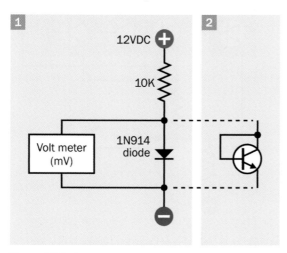

Figure 27-2 *Left, a basic circuit for demonstrating the temperature sensitivity of a diode. Right, an NPN transistor can be substituted to emulate the diode.*

Similarly, the voltage across the p-n junction in an NPN transistor varies with temperature, if the current is constant. A transistor can be substituted for a diode as suggested in section 2 of Figure 27-2. Integrated circuit chips that contain transistors can measure temperature by exploiting this phenomenon.

CMOS Sensors

Some semiconductor temperature sensors use CMOS instead of bipolar transistors. The general concept is similar, but they are described separately, below. See "CMOS Semiconductor Temperature Sensors".

Multiple Transistors

The heat sensitivity of a bipolar transistor can be defined with an equation. If the base-emitter voltage is V_{BE}, q is the charge of an electron, k is a constant (known as Boltzmann's constant), T is the temperature in degrees Kelvin (relative to absolute zero), I_C is the collector current, and I_S is the saturation current (which is less than I_C):

$$V_{BE} = ((k*T) / q) * \log_e (I_C / I_S)$$

The term $\log_e$ means, "the logarithm to base e of the expression in parentheses."

Because k and q have known values, the base-emitter voltage turns out to be proportional to the logarithm of the collector current divided by the saturation current. However, the saturation current depends on the geometry of a transistor, and varies with temperature in a nonlinear way.

To eliminate the factor of saturation current, one transistor can be compared with another transistor that has a larger emitter area. This enables the derivation of a new equation that specifies temperature while getting rid of the troublesome saturation currents, with their nonlinear behavior.

However, it may not be easy to fabricate two transistors, in the same silicon chip, that have the same characteristics except that the emitter area of one is bigger than that of the other. It is much easier to add multiple transistors in parallel, each of them identical to the first. The total emitter area will then be equal to the area in one transistor multiplied by the number of transistors.

In Figure 27-3, assuming all the transistors are at the same temperature in addition to being of identical specification, we can now write two equations. The figure allows room to show only three transistors, but suppose there are N of them. If V_{BE0} is the base-emitter voltage of transistor Q0, on the left, and V_{BEN} is the aggregate base-emitter voltage of the N transistors on the right:

$$V_{BE0} = ((k*T) / q) * \log_e (I_C / I_S)$$

$$V_{BEN} = ((k*T) / q) * \log_e (I_C / N*I_S)$$

From these, an equation can be derived that gets rid of I_C and I_S:

$$V_{BE0} - V_{BEN} = ((k*T) / q) * \log_e (N)$$

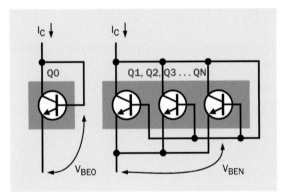

Figure 27-3 *Comparing the base-emitter voltage of one transistor with a set of identical transistors can enable measurement of temperature regardless of the collector current and saturation currents, so long as all the transistors are at the same temperature. See text for details.*

PTAT and the Brokaw Cell

Now if a comparator is added to control the current, a circuit known as the Brokaw Cell is created, shown in Figure 27-4. This is also known generically as a *bandgap temperature sensor*. (A couple of resistors have been omitted for the sake of simplicity.)

Typically, N = 8. That is, there is a set of eight transistors in addition to Q0 (only three being shown here). The voltage difference in the previous equation, V_{BE0} - V_{BEN}, now appears across R2 in the figure, and the voltage across R1 is *proportional to absolute temperature*, often referred to by its acronym, *PTAT*. This voltage can be found from this equation:

$$V_{PTAT} = ((k*T)/q) * \log_e(N) * (2*R1/R2)$$

The Brokaw Cell was the basis of the AD580 chip introduced in 1974 by Analog Devices, and the principle is now used very widely in semiconductor temperature sensors.

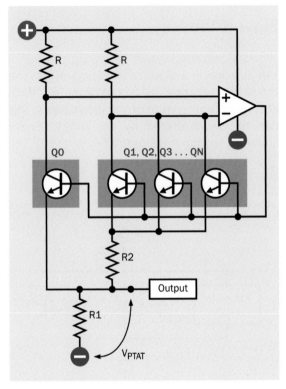

Figure 27-4 *The Brokaw Cell. See text for details.*

Variants

Three output types are used:

- Analog voltage output (voltage varies with temperature).

- Analog current output (current varies with temperature).

- Digital output.

A fourth type creates an output in the form of a square wave, either with a frequency or wavelength proportional to temperature. The Maxim MAX6576 and MAX6577 are examples. However, this type of output is so rare, it is not described in detail here.

Some semiconductor temperature sensors are CMOS-based, and have a voltage output with a negative temperature coefficient. They are described separately. See "CMOS Semiconductor Temperature Sensors".

Analog Voltage Output

LM35 Series

The LM35 is a typical, widely used semiconductor temperature sensor, available from Analog Devices, Texas Instruments, and other manufacturers. Its output voltage changes by 10mV per degree Celsius over a range of approximately -50 degrees to +150 degrees. Accuracy is stated to be plus-or-minus 0.25 degrees at room temperature and plus-or-minus 0.75 degrees over the whole range.

The sensor can be obtained packaged like a transistor, in a TO-92 plastic capsule or metal can. It is also available as a surface-mount component, or in a TO-220 package, like a 5V voltage regulator, as shown in Figure 27-5.

Figure 27-5 *This version of the LM35 can be used to measure surface temperature when secured with a bolt. The background grid is in millimeters.*

This is a three-wire component, two pins or solder pads being used for the power supply while the third serves as the sensor output. The sup-

ply voltage typically ranges from 4V to 30V. Necessary current consumption is only 60µA, which minimizes self-heating.

Because this device is specifically designed for the Celsius temperature range, its output is scaled to 0mV at 0 degrees. A pulldown resistor can be added to measure temperatures below zero.

A bypass resistor of 200 ohms between the output and ground is recommended as a precaution against capacitive effects in the cable run.

The LM34 is almost identical to the LM35, except that its output changes by 10mV per degree Fahrenheit instead of 10mV per degree Celsius.

LM135 Series

Although this sensor contains multiple NPN junctions, the manufacturer describes it as behaving like a zener diode in which the breakdown voltage is directly proportional to absolute temperature. The output increases by 10mV per degree over a range from -55 to +150 degrees Celsius.

For the LM135, the manufacturer claims an error of less than plus-or-minus 1 degree Celsius between 0 and 100 degrees. For the LM235 and LM335, in the same product series, the temperature range is narrower, the accuracy is lower, and the price, also, is lower. An LM335 sensor is shown in Figure 27-6.

Figure 27-6 *A sample of the LM335 temperature sensor in a TO-92 package. The background grid is in millimeters.*

The sensor is available in a TO-92 package (plastic, like a transistor) or a TO-46 (metal can). It is also manufactured in a surface-mount format. The negative terminal is connected directly to ground, while the positive terminal is connected through a series resistor to the positive side of a power supply that can range from 5V to 40V. The third terminal, labeled "ADJ" on datasheets, allows for the output to be adjusted. Figure 27-7 shows the basic circuit. The value of R1 can be chosen to establish an optimal current of 1mA through the sensor, although a range of 400µA to 5mA is tolerable.

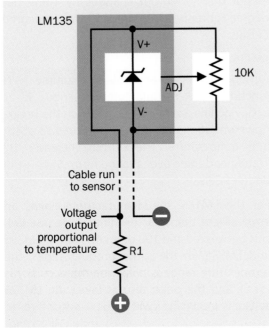

Figure 27-7 Basic schematic for using an LM135, including output adjustment. Because the sensing element behaves like a zener diode, it is represented with the zener symbol.

Analog Current Output

Fewer components exist using output current to measure temperature. The output is applied to a grounded resistor, and the voltage across the resistor then changes with current from the sensor.

The useful aspect of a current output is that its accuracy is unaffected by a cable run as long as 200 or 300 feet. Therefore, this type of component is appropriate as a remote sensor.

LM234-3 Series

This is a three-wire sensor, two wires being used for bias power supply and ground (labeled V+ and V- on the datasheet) and a third (labeled R) that delivers current proportional to temperature. The current from pin R passes through an external resistor to ground, and the voltage across this resistor varies by 214µV per degree Kelvin. A bias voltage ranging from 1V to 40V is required.

A sample of the LM234 is shown in Figure 27-8.

Figure 27-8 *The LM234Z temperature sensor in a TO-92 package. The background grid is in millimeters.*

If the component is used for remote sensing, the resistor should be 230 ohms and can be connected directly between pin R and pin V- of the sensor at the far end of a wire run. At the "home" end of the wire run, temperature output is taken from above a 10K resistor that is placed between the return wire and ground, as shown in Figure 27-9. With these component values, the output voltage will change by 10mV per degree Kelvin.

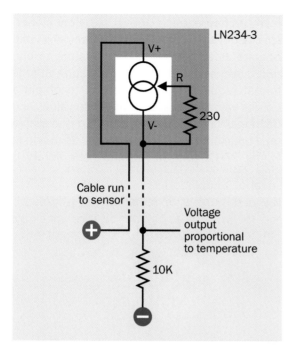

Figure 27-9 *Using an LM243-3 sensor with a current output that varies with temperature.*

The LM234-3 can be encapsulated in a plastic TO-92 package or a TO-46 metal can. A surface-mount version is also available.

The claimed accuracy is plus-or-minus 3 degrees. The temperature range is -25 degrees to +100 degrees.

AD590 Series

The Analog Devices AD590 (successor to the original AD580) is a current-output sensor that uses only two wires. Like the LN234-3 it is available in a TO-46 metal can, but with one lead making no internal connection. It also can be bought in a two-wire "flatpack," or as a surface-mount chip (with eight solder pads, only two of which are connected).

Using a supply voltage of 4V to 30V, the sensor's high-impedance output changes by 1μA per degree Kelvin. Voltage supply variations produce very small errors in the output current; substituting 10V for 5V creates a deviation of only 1μA.

Figure 27-10 shows an application for the AD590 using a resistor and a trimmer to adjust the scale factor. When properly set up, this circuit provides an output that changes by 1mV per degree Kelvin.

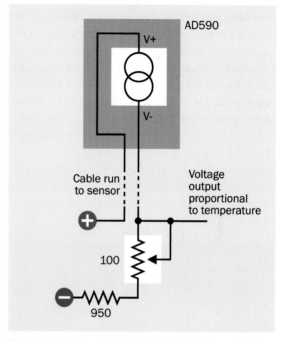

Figure 27-10 *The simplest circuit allowing fine adjustment for use with an AD590 sensor.*

Digital Output

Some of the most popular examples of semiconductor temperature sensors with digital output include the TMP102 series from Texas Instruments, MCP9808 series from Microchip, LM73 series from Texas Instruments and National Semiconductor, and DS18B20 series from Maxim. All of these components can measure a range typical of semiconductor temperature sensors, from approximately -50 to +150 degrees Celsius. Most of them claim an accuracy in the region of plus-or-minus 1 degree over the full range or 0.5 degree in the 0 to 100 range. With the exception of the Maxim DS18B20, which uses its own unique protocol, the components communicate via either the I2C or SMBus protocol.

TMP102 Series

This is available only in a surface-mount format. It has fewer features than the other sensors listed here, and is less accurate, claiming plus-or-minus 3 degrees Celsius over its maximum range of -40 to +125 degrees Celsius. However, it is less expensive. (For greater accuracy, the TMP112 is available.) This is a low-voltage chip requiring 1.4V to 3.6V as its power supply, drawing a quiescent current of 10µA. Temperatures are stored in a 12-bit or 13-bit format that requires some conversion, as a single bit represents 0.0625 degrees Celsius. An alert pin is activated when the measured temperature deviates above or below limits that are preset by the user. No hysteresis adjustment is available for the alerts. The TMP102 is available on a breakout board from Sparkfun, as shown in Figure 27-11.

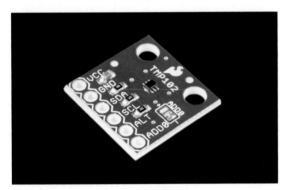

Figure 27-11 *The Texas Instruments TMP102 on a breakout board from Sparkfun.*

MCP9808 Series

This multifunction sensor is available either as a regular surface-mount component, or as a surface-mount with an exposed "thermal pad." It conforms with the I2C bus standard at up to 400kHz, allowing up to 16 sensors to share the same bus. The chip has a variety of temperature alert features, including high and low limits that can activate a dedicated "alert" pin, and a hysteresis value that can be set for the limits, to ignore brief temperature excursions. The chip can be put into "comparator mode," where it simply provides logic-high or logic-low output

if the temperature is above or below a user-specified value. This feature makes the chip operate as a thermostat. Temperature resolution is user-selectable. The temperature storage format requires some conversion to obtain a Celsius value, to deal with negative values and fractional values. However the chip is available on a breakout board from Adafruit, as shown in Figure 27-12, and an Arduino code library is available.

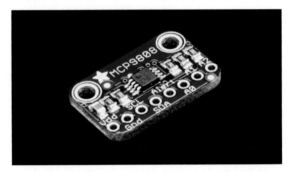

Figure 27-12 *The Microchip MCP9808 on a breakout board from Adafruit.*

LM73 Series

This sensor is only available in surface-mount format. It conforms with the I2C bus standard at up to 400kHz. Its temperature resolution can be set to 11, 12, 13, or 14 bits. An "alert" pin becomes active if the temperature exceeds a preprogrammed limit. An "address" pin can select one of three device addresses by being held in logic-high, ground, or disconnected status. The chip can be put into shutdown mode when power conservation is necessary.

DS18B20 Series

Unlike most digital sensors, this is a three-wire component, because it uses Maxim's proprietary "1-wire bus" with a unique protocol. The bus allows access to a 2-byte register storing digital output from the temperature sensor, but also allows the user to perform other functions, such as setting the resolution of an onboard analog-to-digital converter (which has a maximum resolution of 12 bits), setting a high-temperature and low-temperature alarm, and allowing the sensor to be identified, as each component has a unique 48-bit serial number in ROM.

The chip can draw sufficient power from the data bus to operate, so long as the bus is held high by a 4.7K pullup resistor. (Maxim describes this as "parasite power.") An internal capacitor sustains the chip briefly while the bus is used for its normal purpose of transferring data, but if the bus has low voltage for more than 480µs the chip will reset itself. The "parasite power" feature also will not work above 100 degrees Celsius. Perhaps recognizing that this system may create more problems than it solves, Maxim has also given the component a normal power input pin.

The DS18B20 is available in the TO-92 package and two sizes of surface-mount chip. Its lack of a standard I2C bus, and its use of complicated proprietary codes, create a steep learning curve. Still, it remains a popular sensor, and an Arduino code library for it is available online.

CMOS Semiconductor Temperature Sensors

CMOS variants of semiconductor temperature sensors have appeared relatively recently compared with the bipolar variants. They draw a very low quiescent current (typically, a few microamps) and can work with a power supply from 5.5VDC down to 2.2VDC, making them suitable for handheld battery-powered devices. An analog output is common. Popular examples are the LM20 and the LMT86 series.

Like bipolar sensors, the LMT86 sensors have a limited temperature range, between approximately -50 and +150 degrees Celsius. Again, like the bipolar sensors, they are available optionally in TO-92 and surface-mount packages. A significant difference is that the output has a negative temperature coefficient, diminishing by 10mV per degree Kelvin, because of the characteristics of CMOS semiconductors.

The claimed accuracy is plus-or-minus 0.25 degrees Celsius. The output voltage covers a range of about 2V, diminishing from 0.5V below the supply voltage at -50 degrees Celsius.

A sample of the LMT86 is shown in Figure 27-13.

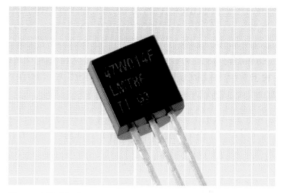

Figure 27-13 *A sample of the LMT86 CMOS temperature sensor. The background grid is in millimeters.*

What Can Go Wrong

Different Temperature Scales

Some voltage-output sensors create an output convertible to degrees Kelvin, while others use degrees Celsius. While the temperature degrees are the same in each scale, a component with an output in millivolts may assign 0mV either to 0 degrees Celsius or 0 degrees Kelvin (equivalent to -273.15 degrees Celsius). The advantage of using a Kelvin scale is that it avoids the problem of negative temperature values.

Rarely, a sensor may use degrees Fahrenheit.

Interference in Cable Runs

Sensors with a voltage output are susceptible to electrical interference. Twisted-pair or shielded cable runs are recommended when sensors are placed remotely.

For the Maxim DS18B20, which uses a 1-wire bus, multiple sensors should be connected along one run of wire (linear topology) instead of each sensor being connected to a central point (star topology). If the cable lengths are longer than a few meters, the topology starts to matter.

Latency

The packaging of semiconductor temperature sensors can create latency in their response time. While a thermocouple consists only of a pair of wires joined by a small dot of melted metal, a TO-92 semiconductor package adds thermal mass that will slow the response considerably. Moreover, copper leads will conduct heat from a circuit board if the board is warmer than its environment.

Surface-mount chips have a very low mass, but must be soldered to some kind of board, even if it is a very small one.

Generally, other types of sensors may be appropriate where rapid response is necessary.

Processing Time

In a sensor with a digital output, the onboard analog-to-digital converter will add a small delay before the data becomes available, and during that delay, the component cannot respond to a new temperature. The output from an analog device may be more suitable for rapid detection of temperature variations.

infrared temperature sensor

An **infrared temperature** sensor is sometimes described as a *thermopile*. In reality the sensor module *contains* a thermopile. In this Encyclopedia, a thermopile is considered to be a separate component, described in the entry discussing **thermocouples**. See "Thermopile" in Chapter 25.

Other terms that are sometimes used for an infrared temperature sensor are *contactless thermometer* or *infrared thermometer*. This Encyclopedia classifies a thermometer as a commercially marketed product, not a component.

Devices such as a *radiation pyrometer, IR pyrometer, optical pyrometer,* or *thermal imager* provide ways of measuring infrared radiation, but are outside the scope of this Encyclopedia.

A **passive infrared** motion sensor (PIR) can detect infrared radiation, but only responds to fluctuations in intensity. An infrared temperature sensor measures the steady-state value of incident radiation.

OTHER RELATED COMPONENTS

- **passive infrared** motion sensor (see Chapter 4)
- **thermocouple** (see Chapter 25)

What It Does

Most temperature sensors discussed in other entries in this Encyclopedia are *contact sensors*, meaning that to measure the temperature of an object, liquid, or gas, they must make contact with it. In situations where contact is not possible or desirable, an **infrared temperature** sensor can be used. It responds to the *black-body radiation* (sometimes known as *characteristic radiation*) that is emitted by all materials above absolute zero (0 degrees Kelvin). This varies with temperature as a result of the movements of molecules.

Situations where noncontact sensors may be preferable to contact sensors include:

- An object is inconveniently located or too far away.
- The temperature of a large area must be measured.
- Contact with a small object would change the temperature of the object. The act of measurement would change the value being measured.
- The object is corrosive, abrasive, or otherwise liable to damage a sensor.
- The object is moving or vibrating.
- The surface of the object must not be contaminated (for example, unprotected foods).

- The temperature of the object is lower than around -50 degrees Celsius or higher than 1,300 degrees Celsius.

However, noncontact sensors have some limitations:

- Normally, only the surface temperature of a target can be measured.

- The optics of the sensor must be protected from dust, dirt, and liquid.

- The target must be clearly visible, in line-of-sight.

- Air pollution will degrade the temperature measurement. Some gases, such as carbon dioxide, will tend to absorb infrared radiation.

- The sensor will be affected by other sources, including reflected, transmitted, and convective heat.

- While an infrared sensor can theoretically respond to a very wide range of temperatures, in practice separate sensors of differing sensitivity are needed to cover a full range.

- Different types of materials emit differing intensities of black-body radiation, even if they are at the same temperature. Some compensation is necessary, or the surface of the object may have to be painted.

Applications

Handheld contactless thermometers were an early application for noncontact sensors.

In astronomy, the thermal radiation from the Sun and other stars is of interest to astronomers.

More recently, the declining cost of an infrared temperature sensor, and the ease of deploying it, have made it appropriate in consumer products. A significant area of adoption is in notebook computers and handheld devices, where

processor performance must be balanced against the need to prevent the case from becoming too hot to hold comfortably. In this kind of application, gluing a sensor to the interior of the case would be a manual operation during the production process and would require a wired connection. An infrared temperature sensor mounted on the circuit board, viewing the underside of the case, can achieve the same objective more simply.

A contactless sensor is also very useful for measuring the temperature of rotating objects, such as heating rollers in a laser printer.

Schematic Symbol

No specific schematic symbol exists for an infrared temperature sensor.

How It Works

While nanometers (abbreviated nm) are generally used to measure visible wavelengths, the longer wavelengths of far-infrared are often measured in micrometers (abbreviated μm). The measurable infrared values are defined as ranging from 0.7μm to 14μm, corresponding to peak emissions from a black body ranging in temperature from 200 degrees Kelvin to 6,000 degrees Kelvin (about -70 to +5,700 Celsius).

Unfortunately an object does not emit just one wavelength of black-body radiation for each temperature value. It emits a spread of wavelengths that becomes wider as the temperature increases. However, the peak intensity also increases with temperature, when measured as *spectral radiance*, which is defined as watts per steradian, per micrometer of wavelength. (A steradian is the solid angle at the top of a cone, in this case the cone being of emitted radiation.) Because the intensity increases, it can be used to calculate the temperature.

Figure 28-1 illustrates this concept. Note that both of the axes have logarithmic scales.

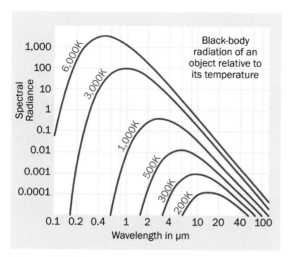

Figure 28-1 *The increase in intensity, and the widening spread of wavelengths, of black-body radiation emitted by one object at six different temperatures (in degrees Kelvin).*

All the curves are for one object. Each curve is specific to one temperature in degrees Kelvin, showing how the intensity of emitted radiation varies with wavelength. Note that radiation at wavelengths shorter than 0.7µm is within the visible spectrum; thus, objects at 1,000 degrees Kelvin, or hotter, may be seen to glow visibly.

Because of the wide variety of intensities and temperatures, an infrared sensor that is ideal for measuring a temperature of 1,000 degrees Kelvin will not provide an accurate result at 200 degrees Kelvin. The peak spectral radiance at 1,000 degrees is more than 10,000 times the peak at 200 degrees. Also, the curves in the figure are for an "ideal" object emitting pure black-body radiation. In reality, glass, plastic, and many other materials have a much lower *emissivity*, meaning that they emit less radiation, and are classified as *gray bodies*. A metallic object with a polished surface may emit one-tenth of true black-body radiation.

These issues cannot be ignored, but may be dealt with by relatively simple strategies. Infrared sensors will be rated according to their suitability for different temperature ranges, and the emissivity of the object being measured

can be determined by consulting standard tables. Alternatively, the object can be spray-painted with special black paint (such as "Senotherm" or "3-M Black") that has a known emissivity of about 0.95 of pure black-body radiation. Alternatively, a specially formulated black sticker may be applied to the object that is being measured, so long as its temperature is within reasonable limits.

However, a basic infrared temperature sensor will not function reliably if it is pointed randomly at a variety of objects that vary widely in temperature. Specialized, expensive industrial devices incorporate compensation to deal with these issues, but they are outside the scope of this Encyclopedia.

Thermopile

A typical low-cost, chip-based infrared temperature sensor contains a *thermopile* consisting of multiple thermocouples etched into silicon and connected in series. The concept of a thermopile is illustrated in Figure 25-9, where a brief explanation is included.

The configuration of the thermopile is arranged so that the hot junctions of its thermocouples are all clustered in a small central area, where they receive incoming radiation through a window (often made of silicon) that is transparent to infrared wavelengths. The cold junctions are dispersed around the periphery, where they are shielded from incoming radiation. One way to visualize this is shown in Figure 28-2, although this is not a literal depiction of an actual sensor.

Instead of using alternating types of wire, as in a **thermocouple**, a chip-based thermopile often uses alternating segments of n-type and p-type silicon. The hot junctions are mounted on a thin film that has very little heat capacity, while the cold junctions are mounted on a thicker substrate that acts as a heat sink.

Figure 28-2 *Simplified diagram of the thermopile configuration inside an infrared sensor chip. Radiation arriving through a window in the chip affects the thermocouple junctions in the central region, while junctions around the edges remain at a lower temperature.*

Temperature Measurement

The voltage generated by the thermopile is related to the difference in temperature between the hot and cold thermocouple junctions. Thus there are three interrelated variables: hot temperature, cold temperature, and voltage. To calculate one variable, we must know the other two.

The hot temperature is what we wish to know. Therefore we must establish the voltage (which can easily be measured) and the cold temperature. The cold temperature can be determined by adding a thermistor inside the chip.

Typically an infrared temperature sensor with an analog output will have two pins that provide access to the internal thermistor, so that its temperature can be calculated from its resistance. Another two pins provide the voltage between the ends of the thermopile.

Interpreting and reconciling these values is not a trivial matter, especially bearing in mind that

the thermistor has a negative temperature coefficient and a nonlinear output, and the thermopile will also have some nonlinearity. To simplify this situation, some infrared temperature sensors incorporate electronics to perform the necessary calculations and provide a digital output. This output can be converted to degrees of temperature by some fairly simple mathematical operations in an external microprocessor.

Variants

Two types of sensors are popular. One is surface-mounted, such as the TMP006, shown in Figure 28-3. This type generally has a digital output. The other type is a discrete component with four leads, such as the Amphenol ZTP135, shown in Figure 28-4. Discrete components may have either an analog or a digital output.

Figure 28-3 *A surface-mount infrared temperature sensor with digital output. Eight tiny solder pads are located on the underside. The background grid is in millimeters.*

Both types of sensors allow infrared light to enter through an area that is opaque to the visible spectrum but transparent to the appropriate range of wavelengths.

Figure 28-4 *A through-hole infrared temperature sensor with analog output. The background grid is in millimeters.*

The ZTP135 has an analog output shown in Figure 28-5.

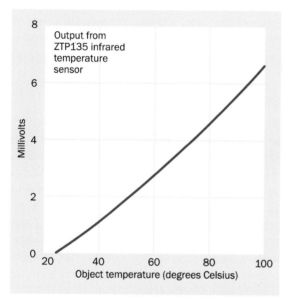

Figure 28-5 *Analog output from an infrared temperature sensor.*

The TMP006 is only about 1.5mm square, but is available on a breakout board from Sparkfun. Its successor, the TMP007, is available on a breakout board from Adafruit, shown in Figure 28-6.

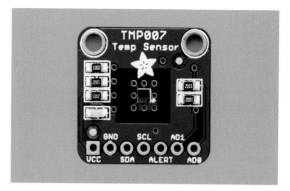

Figure 28-6 *The TMP007 sensor mounted on a breakout board from Adafruit.*

Surface-Mount Specifications

The TMP006 and TMP007 require a supply voltage that can range from 3.3V to 5V. These chips support the SMBus and I2C bus protocols, using a bus address that can be user-selected. An internal analog-to-digital converter uses 1 least-significant bit to represent 1/32nd of a degree Celsius, and data is saved as a 14-bit signed integer. Up to 16 temperature samples can be averaged internally.

Measurable temperature range is -40 to +125 degrees Celsius. Some hysteresis is built in. The TMP007 supports an alert mode if temperature falls above or below a user-specified threshold.

Sensor Arrays

Using multiple thermopile sensors arrayed in a line or a grid, with an array of lenses, it is possible to capture an image of temperature variations over a surface or a scene. This is known as *thermal imaging*. It can detect heat leakage from buildings, indicating poor insulation, or can locate hot spots in electronic circuits. Heimann Sensor has pioneered the miniaturization of a 31 x 31 grid of thermopile sensors in a single TO-8 or TO-39 package.

Values

Temperature Range

Chip-based infrared temperature sensors are typically designed for a range from about -20 degrees Celsius to about +125 degrees Celsius. Their peak sensitivity is to wavelengths between 4μ and 16μm.

Other types of infrared temperature sensors can have a much wider temperature range, but are more costly.

Field of View

Often referred to by its acronym, *FOV*, the field of view is the angle at the apex of an imaginary cone extending outward from the sensor, defining a boundary where the sensitivity diminishes below 50% of the value directly in front of the sensor. Greek letter φ may be used to represent the angle between the surface of the cone and the center line, while θ represents the angle between the opposite surfaces of the cone (i.e., 2 * φ). This is shown in Figure 28-7. θ is usually the angle defined as field of view.

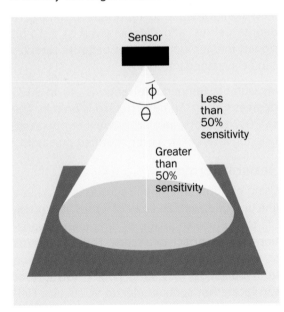

Figure 28-7 *Measurement of the field of view from a sensor, defined as the boundary of an imaginary cone where sensitivity drops below 50%.*

Because most infrared temperature sensing components do not have a lens, they are sensitive over a wide angle. The field of view is typically 90 degrees.

What Can Go Wrong

Inappropriate Field of View

The object being evaluated must fill the field of view of the sensor, to avoid measuring other objects around it.

Reflective Objects

A reflective object has lower infrared emissivity, and also may provide misleading output if the sensor is actually measuring thermal radiation reflected in the surface of the object in front of it. In a permanent installation, such as inside a device, the surface to be measured may have to be painted to reduce its reflectivity.

Glass Obstruction

Because glass is opaque to the spectrum of infrared that is of interest, temperature cannot be measured through a glass window. Silicon is opaque to visible wavelengths but is transparent to wavelengths longer than 2μm.

Multiple Heat Sources

Heat is transferred by convection, conduction, and radiation. While an infrared temperature sensor is designed to be sensitive to radiation, it will also respond to other heat sources. Warm or cold air currents will affect its response, and so will heat conducted through the material on which it is mounted. Careful placement of the sensor is important. A shield around the sensor, with a small hole in the center, can prevent convection, while correct location on a circuit board can minimize conduction.

Thermal Gradients

An infrared temperature sensor should be mounted in a stable environment where it will not be exposed to thermal gradients (one side being hotter than another). This asymmetry can cause inaccurate readings.

microphone

29

OTHER RELATED COMPONENTS

- **speaker** (see Volume 2)
- **headphone** (see Volume 2)

What It Does

The sensation of sound is created by rapid waves of air pressure impinging upon the eardrum. A microphone can convert these pressure waves into an alternating electrical signal that can be amplified, recorded, broadcast, transmitted through wires, and reproduced as sound by a headphone or speaker. The principle is illustrated in Figure 29-1. (For more information about sound reproduction, see the entries on **headphone** and **speaker** in Vol. 2.)

Schematic Symbol

Various schematic symbols for a microphone have been used during the decades since its invention. A selection is shown in Figure 29-2. Each symbol assumes that sound is traveling from left to right. This is important when interpreting the symbol at top right, which can represent an earphone when it points in the opposite direction. Unfortunately, some schematics do not conform with this rule.

The two symbols at the bottom, showing a capacitor inside the microphone, should be reserved for condenser or electret microphones.

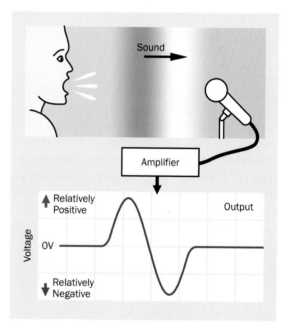

Figure 29-1 *The principle of converting pressure waves into an alternating electrical signal (adapted from an illustration in* Make: More Electronics*).*

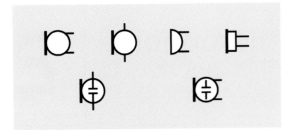

Figure 29-2 *A selection of schematic symbols that represent a microphone.*

How It Works

Some types of microphones generate a small voltage, while others have a fluctuating resistance that modulates a DC current.

Carbon Microphone

This was a very early attempt to reproduce sound. It contained carbon granules whose packing density increased and decreased in response to air pressure waves. When the density increased, the resistance between the granules diminished, and vice-versa. The principle is illustrated in Figure 29-3, and was patented by Thomas Edison in 1877 for use in telephones. As late as the 1950s (and even later in some countries), wired telephone handsets contained carbon microphones. Their bandwidth was extremely limited.

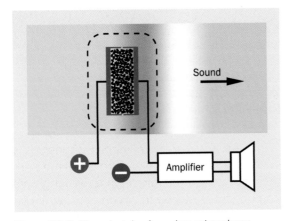

Figure 29-3 *The principle of a carbon microphone.*

Moving-Coil Microphone

Also known as a *dynamic* microphone, this consists of a very small, light coil of thin wire on a cylindrical tube that can vibrate along the axis of a permanent magnet. This principle is illustrated in Figure 29-4. A diaphragm is attached to the front of the tube, and responds to air pressure waves that penetrate the perforated enclosure of the microphone. Movements of the coil around the magnet create small alternating currents in the wire. The inertia of the coil, tube, and diaphragm, and the force needed to overcome the interaction between the coil and the magnet, impose a limit on the high-frequency response of this design.

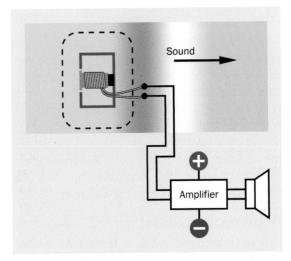

Figure 29-4 *The principle of a moving-coil microphone.*

Condenser Microphone

This type of microphone contains two thin discs or plates that form a capacitor. (In the early days of electricity, a capacitor used to be known as a condenser. The terminology has persisted for microphones.) An equal and opposite charge is applied to the plates. One plate is flexible, and as it responds to pressure waves, the capacitance between it and the other, rigid plate fluctuates. If the charge on the plates is kept approximately constant while the capacitance fluctuates, the voltage across the capacitor fluctuates also. These fluctuations can be amplified, as suggested in Figure 29-5.

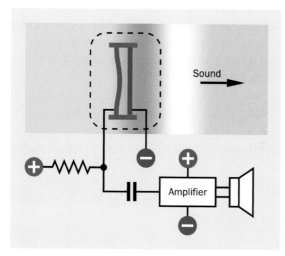

Figure 29-5 *The principle of a condenser microphone.*

Electret Microphone

This works on the same principle as a condenser microphone, except that its plates are made from a ferroelectric material that retains an electrical charge, just as iron will retain a magnetic polarization. The name of the microphone is derived from "electrostatic" and "magnet." While early electret microphones were of poor quality, they have evolved to rival condenser microphones, and are extremely affordable. Because the electret creates very small currents, it usually includes a transistor or op-amp in its package to boost the signal, and has an open-collector output. The basic circuit for an electret is shown in Figure 29-6. For more information about using an open-collector output, see Figure A-4.

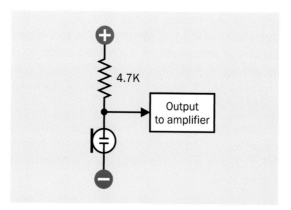

Figure 29-6 *The basic circuit for using an electret microphone.*

An example of a low-cost electret microphone is shown in Figure 29-7. This type of component is sold either with leads attached, or with solder pads.

Figure 29-7 *A generic electret microphone. The background grid is in millimeters.*

MEMS Microphone

This type, often used in mobile phones, is a capacitive device that works on the same principle as a condenser microphone, although the component is etched in silicon and has a diaphragm that measures only about 1mm square. Many MEMS microphones have an analog output that is amplified in the same chip. Others have a digital output, using *PDM encoding*. This reduces the analog signal to a very fast bit stream, in which the density of the bits represents the amplitude of each fluctuation in a sound wave. PDM is an acronym for *pulse density modulation*. It requires an external clock signal to time the bit stream.

A breakout board from Sparkfun, on which is mounted an Analog Devices ADMP401 MEMS microphone with a preamplifier, is shown in Figure 29-8.

Figure 29-8 *A breakout board for a MEMS microphone (the metal-clad rectangular component at the far end).*

Piezoelectric Microphone

This has also been known as a *crystal* microphone. It contains a diaphragm that functions as a transducer. When it flexes in response to pressure waves, the mechanical energy is transformed into a small amount of electrical energy. Piezoelectric microphones were replaced by the moving-coil type in domestic audio devices when vacuum tubes were replaced with transistors, but may still be used as contact microphones to amplify acoustic musical instruments, or to trigger the playback of digitally sampled musical sounds.

Other variants include *ribbon* microphones (which were common in recording studios in the 1950s and 1960s, but have become rare), *laser* microphones, and *fiber-optic* microphones. They are not sufficiently common to be included in this entry.

Values

Sensitivity

Sound pressure is a complicated topic, explored in detail in the **transducer** entry in Volume 2. It can be measured in pascals, where 1 pascal = 1 newton per square meter.

The *sound pressure level* is a different concept. It measures the *relative* intensity of a sound, in a logarithmic scale calibrated in *decibels* (abbreviated dB). The reference value for this relative scale is 20 micropascals, considered to be the threshold of human hearing, comparable to a mosquito three meters away. This is assigned the value of 0dB.

From this point upward, the actual sound pressure doubles for each additional 6dB. A table of noise sources and their approximate decibel values is shown in Figure 29-9. This is derived from averaging eight similar tables, which are not always consistent in their estimates. It is an approximate guide only.

Decibels	Noise Example
140	Jet engine at 50 meters
130	Threshold of pain
120	Loud rock concert
110	Automobile horn at 1 meter
100	Jackhammer at 1 meter
90	Propeller plane 300 meters above
80	Freight train at 15 meters
70	Vacuum cleaner
60	Business office
50	Conversation
40	Library
30	Quiet bedroom
20	Leaves rustling
10	Calm breathing at 1 meter
0	Auditory threshold

Figure 29-9 *Decibel values for some common sound sources. Reproduced from Volume 2 of this Encyclopedia.*

The decibel unit is important when understanding the specifications of microphones, because it is used to measure their response. Microphone sensitivity is established with a standard input sine wave of 1kHz in frequency and 94dB in intensity (equivalent to 1 pascal in actual sound pressure), measured at the microphone. The sensitivity of an analog microphone is then defined as the number of decibels in an output signal of 1V. Because the output is an AC signal, voltage is measured as a root-mean-square (RMS) value.

For digital microphones, sensitivity is measured as the decibels that can be reproduced by a full-scale digital output. This value is abbreviated as dBFS.

Directionality

A microphone that has a directional response is desirable in many situations. Often, for example, sounds in front of the microphone are more important than sounds from behind the microphone. The directionality of a microphone (sometimes referred to as its *directivity*) is usually represented with a *polar graph* in which the microphone is seen from above, and its sensitivity to sounds from various directions is shown with a curve such as those in Figure 29-10. The circles are drawn at intervals of 5dB, with 0dB at the periphery and -30dB at the center. The precise response for an individual microphone should be shown in its documentation.

Frequency Response

Every microphone tends to be more sensitive to some sound frequencies than others. A manufacturer will provide a graph showing this sensitivity, in decibels, plotted against sound frequency on a logarithmic horizontal axis. Theoretically, human hearing extends from around 20Hz to 20kHz, but few people are actually capable of hearing the high end of that range, and 15kHz may be a more realistic limit for a young person, diminishing to 10kHz with middle age.

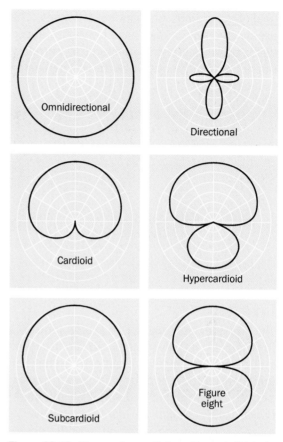

Figure 29-10 *Six generic sensitivity patterns. Individual microphones will deviate somewhat from these generic curves.*

An ideal flat response would show that a microphone is equally sensitive to all frequencies. In reality, *rolloff* usually occurs at low frequencies, and will eventually occur at high frequencies, although it may be preceded by a rise in response. If the central section of a curve is flat within plus-or-minus 1dB, this is a level of performance that was attained only by expensive studio microphones in the past. Electrets and MEMS microphones can now provide equivalent frequency response for $1 or $2 apiece, as opposed to the hundreds or even thousands of dollars that used to be necessary for professional equipment.

The response curve shown in Figure 29-11 is for the eMerging i436, an electret microphone sold in a module as an accessory to enable high-quality recordings on handheld devices. The rise around 15kHz may have been introduced deliberately by the manufacturer to compensate for reduced sensitivity of the human ear in that range.

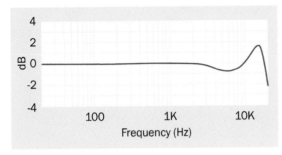

Figure 29-11 *Frequency response for an electret microphone.*

Impedance

The impedance value for a microphone is a function of its resistance, capacitance, and inductance. An amplifier input will also have an impedance rating, and for ideal power transfer between microphone and amplifier, the impedance values should be identical. However, a more important consideration in audio equipment is to avoid voltage loss between the output device (in this case, the microphone) and the input device (the amplifier). To achieve this, the output device should have a low impedance, while the input device should have a high impedance. Most microphones are rated at 150 to 200 ohms, while an amplifier may be rated at 1.5K to 3K.

Total Harmonic Distortion

When an audible sine wave is converted to an electrical output by a microphone (and by its preamplifier, if one is included in the module), the output may become corrupted by some multiples of the basic frequency. These are known as *harmonics*, and they are considered as a distortion of the signal. Total harmonic distortion, measured by a spectrum analyzer over the entire frequency range, should ideally be less than 0.01%.

Signal-to-Noise Ratio

Often abbreviated as S/N or SNR, the signal-to-noise ratio in a microphone is measured in decibels, and should be 60dB or higher.

What Can Go Wrong

Cable Sensitivity

Audio amplification is always vulnerable to electrical noise, which tends to be amplified along with the signal. Small signals from microphones require the use of shielded cables to reduce hum and other types of interference.

Noisy Power Supply

For similar reasons, power supplies must be as free as possible from voltage spikes and other fluctuations.

current sensor

30

The entry describes components that can be installed to monitor current on an indefinite basis. It does not include test equipment, such as test meters or multimeters.

A *current transformer* may be used to measure current, but is not included in this Encyclopedia.

OTHER RELATED COMPONENTS

- **voltage** sensor (see Chapter 31)

What It Does

A current sensor measures the flow of electricity through a wire or a device, and supplies an output that can be interpreted either visually or by a microcontroller to provide a reading in amperes or fractions of an ampere.

Applications

Current sensing is important in industrial applications such as the control of high-powered motors. It can be used to monitor the performance of an inverter, or for everyday purposes such as monitoring the long-term power consumption of an appliance. During product development, a current sensor can indicate the power consumption of a circuit as it changes with modifications.

This entry describes three methods to measure current: an ammeter, series resistor, and Hall sensor. While other methods exist, they are outside the scope of this Encyclopedia.

Ammeter

An ammeter that is sold as a standalone device with leads for circuit testing is often described as a *test meter*. Its functionality is usually built into a *multimeter*. Test meters and multimeters are outside the scope of this Encyclopedia.

An ammeter designed for permanent installation in a device or prototype is a type of *panel meter*, such as the one shown in Figure 30-1. This traditional-style analog meter may be less expensive than the many digital types that are available. It uses a magnetic field created by current flowing through a coil to pull a needle across a scale, against the force of a spring.

Figure 30-1 *A traditional-style analog ammeter.*

A digital ammeter allows a wider range of values to be viewed more easily. The meter from Adafruit in Figure 30-2 has a range of 0A to 9.99A, at voltages from 4.5VDC to 30VDC. The meter can be powered parasitically from the currents that it is measuring, or can use a separate isolated 5VDC supply.

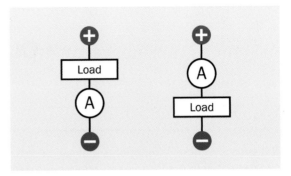

Figure 30-4 *Two options for placement of an ammeter in a circuit.*

However, regardless of the placement of the meter, the process of measuring current will inevitably change the value of the current being measured. This is because the ammeter imposes some internal resistance of its own. The resistance is extremely low, and may be considered negligible for loads of more than a few ohms.

- The low internal resistance of an ammeter means that it must never be connected in parallel with a load, or directly across a power source.

A disadvantage shared by analog and digital meters is that they are not usually interchangeable between AC and DC.

Figure 30-2 *A panel-mount digital ammeter that can measure up to 9.99A.*

Schematic Symbol

An ammeter may be represented in a schematic with the letter A inside a circle, as shown in Figure 30-3.

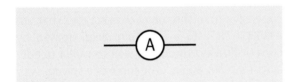

Figure 30-3 *An ammeter may be represented like this in a schematic.*

Ammeter Wiring

Two ways to use an ammeter in a circuit are illustrated in Figure 30-4, where the load may be any equipment, device, or component that provides some electrical resistance. Because current is the same at all points in a simple circuit, the current that the meter measures flowing through itself will be the same as the current flowing through the load, and the sequence of components is immaterial.

Series Resistor

The current flowing through a load can be calculated by measuring the voltage across a series resistor that is inserted between the load and its ground connection. The concept is illustrated in Figure 30-5.

Using Ohm's law, if U is the voltage drop, I is the current, and R is the value of the resistor:

$$I = U / R$$

This tells us that for a fixed value of R, current is proportional to voltage. Therefore, measuring the voltage enables calculation of the current, so long as the value of R is known.

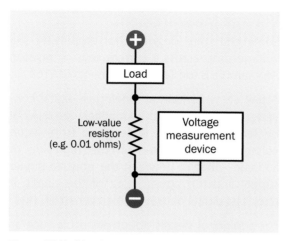

Figure 30-5 *A basic circuit for measuring current as a function of the voltage drop across a small-value resistor. The load in this figure is any circuit or device with a relatively higher resistance. The voltage measurement device could be a microcontroller or analog-to-digital converter.*

Suppose that R has a very small value, which is trivial compared with the resistance of the load. Consequently, the current in Figure 30-5 will be mainly determined by the load, and we may consider the value of the current to be almost the same with or without the addition of R. In that case, the voltage drop across the resistor will be smaller if R is smaller. A smaller voltage drop will not be as easy to measure, but a lower resistance will result in less wastage of power.

If P is the power:

$$P = R * I^2$$

An example may help to make this clear. Suppose the resistor has a value of 0.5 ohms, and the voltage drop across it is measured to be 1V. Ohm's law shows that current flow is 1 / 0.5 = 2A. The power formula shows that P = 0.5 * 4 = 2W.

To waste less power, the value of the resistor should be reduced further. Suppose a resistor of 0.01 ohms is used, and the voltage drop across it is measured as 0.02 volts. The current is 0.02 / 0.01 = 2A, as in the preceding example, but the power dissipation is now only 0.01 * 4 = 0.04W, which is negligible.

But are resistors available that have values measured in fractions of an ohm?

Current-Sense Resistors

In fact, many *current sense* resistors are available, with values of 0.1 ohms, 0.001 ohms, 0.0001 ohms, and many in between. Some resistors have values measured in micro-ohms. Examples are shown in Figures 30-6, 30-7, and 30-8.

Measuring a small voltage drop is easily done by using a microcontroller. However, the connection to the microcontroller must be made as near to the resistor as possible, to eliminate the additional resistance of wires or circuit-board traces. For this reason, precision current-sense resistors may be equipped with four terminals. Two are wider, and are meant for connection to the flow of current. The other two are narrow, for measuring the voltage over the resistor. With this *4-point* configuration, voltage drop over the resistor can be measured as close to it as possible. The 0.001-ohm surface-mount resistor in Figure 30-8 is designed for 4-point measurement.

Figure 30-6 *Two current resistors manufactured by KOA Speer. Left: 0.1 ohms, 5W, 5%. Right: 1 ohm, 5W, 5%. The background grid is in millimeters.*

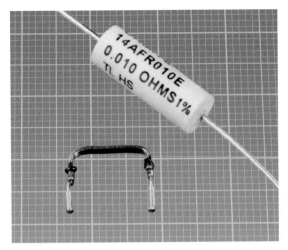

Figure 30-7 *Two styles of current resistors rated for 0.01 ohms. Bottom left: A plug-in version from TT Electronics, rated 1W and 5%. Top right: Ohmite, 4W, 1%. The background grid is in millimeters.*

Resistors that have the lowest values and are intended to tolerate high current may consist of just a metallic strip welded to solderable pins. This type of component is sometimes called an *open-air resistor*. It is commonly used in multimeters, for measuring currents up to 10A or above.

Figure 30-8 *This Vishay 4-terminal surface-mount current resistor is rated for 0.001 ohms, 3W, 1%. The background grid is in millimeters.*

Voltage Measurement

Some chips are designed for amplifying the voltage drop across a current-sensing resistor. An example is the Texas Instruments INA169.

A few chips contain an analog-to-digital converter in addition to the amplifier. The INA219 by Texas Instruments is designed to measure voltage as well as current, on the "high side" of a circuit—that is, between the positive power supply and the power input of the circuit. It makes its digital data available over an I2C bus.

For additional details about protocols such as I2C, see Appendix A.

Measuring current from the voltage drop across a series resistor offers the advantages of simplicity, ability to work with AC or DC, and low cost (although some resistors of extremely low value can be relatively expensive). A possible disadvantage is that the measurement circuit is not isolated from the circuit whose current is being measured.

Hall-Effect Current Sensing

The principle of a Hall-effect sensor is explained in the entry for **object presence** sensors. See "Hall-Effect Sensor". Normally this type of sensor is activated by an external permanent magnet, but it can also react to the magnetic field generated by current flowing through a wire.

Because the field generated around a wire is proportional with the current, the analog output voltage generated by a linear Hall-effect sensor can also be proportional with the current.

Hall-effect sensors for this specific purpose are available in 8-pad surface-mount packaging. The current to be measured passes through a copper conductor that is embedded in the chip. An example is Allegro's ACS712, for AC or DC currents up to 30A. The internal resistance of the current path through this chip is stated as 1.2 milliohms, and the path is isolated from the sensing circuitry.

Three variants of the chip are available, for currents of plus-or-minus 5A, 20A, and 30A. Depending on which variant is used, the output will range from 66mV to 185mV for each increase of 1A in the current path. Because the current path is isolated, the chip requires a separate power supply of 5VDC.

The 5A version of the AVS712 can be bought on breakout boards from Sparkfun. In Figure 30-9, the board at top-left contains only the ACS712, while the board at lower-right adds an op-amp with sensitivity control, to amplify the voltage output when measuring small currents.

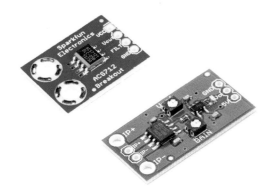

Figure 30-9 *Breakout boards using the AVS712 Hall-effect current sensor. The one at lower-right includes an op-amp to amplify small signals.*

What Can Go Wrong

Confusing AC with DC

A panel meter that is designed only to measure DC should not be used with AC, and vice-versa.

Erroneous readings or damage to the meter may result.

Magnetic Interference

A disadvantage of Hall-effect current sensing is that the sensor can be affected by stray magnetic fields. Because a Hall-effect chip is responding to very small magnetic effects, it is vulnerable to interference. Consult the manufacturer's datasheet carefully regarding correct placement of a chip on a circuit board.

Incorrect Meter Wiring

The correct wiring of an ammeter is in series with a load, not in parallel with a load. This may seem an elementary error, but is easy to make if an ammeter and a volt meter are both being used, and they look similar.

Because some panel meters are not fused, applying an ammeter directly across a power supply, without any series resistance, may result in immediate and hazardous destruction of the meter.

Incorrect wiring can also occur when using a digital meter that has four wires emerging from it: two for current testing and two for a separate power supply. This issue is especially important when measuring substantial currents (above 1A).

Current Out of Range

Attempting to measure a current that exceeds the range of an ammeter may damage the meter or blow its internal fuse, if it has one.

voltage sensor

31

The entry describes the type of component that can be installed to monitor voltage on an indefinite basis. It does not include test equipment, such as test meters or multimeters.

OTHER RELATED COMPONENTS

- **current** sensor (see Chapter 30)

What It Does

A voltage sensor measures the electrical potential between any two points in a circuit, or the voltage supplied by a power source, and provides data in volts or fractions of a volt. It should not be confused with an *analog-to-digital converter*, which is not a sensor in itself, but can process the voltage output from a sensor by digitizing it. More information about analog-to-digital conversion is in the Appendix. See Appendix A.

Applications

Voltage measurement is important in conjunction with all types of power supplies, to verify their performance. A volt meter may also be used to show the output from various types of analog sensors that have voltage output.

A graphical display may be used in audio equipment to indicate signal level, which can be proportional with voltage.

Volt Meter

A volt meter that is sold as a standalone device with leads for circuit testing is often described as a *test meter*. Its functionality is usually built

into a *multimeter*. Test meters and multimeters are outside the scope of this Encyclopedia.

A volt meter designed for permanent installation in a device or prototype is a type of *panel meter*, which is described here.

An antique analog panel meter is shown in Figure 31-1.

Figure 31-1 *An antique analog volt meter.*

The four scales on the dial correspond with separate input terminals at the rear of the unit.

The unequally divided scales are a simple way to compensate for the nonlinear response of the mechanical movement inside the meter.

Modern analog volt meters are still manufactured, and may be cheaper than their digital equivalents. However, a digital volt meter allows a wider range of values to be viewed more easily. The meter in Figure 31-2, sold as a low-cost battery tester, measures voltages from 4VDC to 13VDC to an accuracy of two decimal places. It needs no separate power supply.

Figure 31-2 *A panel-mount digital volt meter.*

Sometimes the number of digits in a volt meter is specified as ending with one-half, as in 3.5 or 3-1/2 digits. This means that the most significant (leftmost) digit can only be a 1 or a blank. The extra "half digit" may seem inconsequential but doubles the range of displayable values. For example, a 2-digit display can only show 100 values, from 0 to 99. A 2-1/2-digit display can show 200 values, from 0 to 199.

Schematic Symbol

A volt meter may be represented in a schematic with the letter V inside a circle, as shown in Figure 31-3.

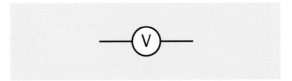

Figure 31-3 *A volt meter may be represented like this in a schematic.*

Volt Meter Wiring

Two ways to use a volt meter in a circuit are illustrated in Figure 31-4, where the load may be any equipment, device, or component that provides some electrical resistance.

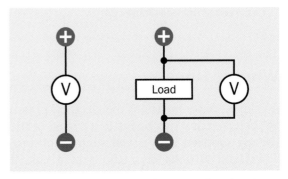

Figure 31-4 *Two options for placement of a volt meter.*

On the left side of the figure, a volt meter may be connected directly across a power source, because the internal resistance of the meter is so high, it will draw very little current. In this configuration, the meter measures the voltage of a source when the source is virtually unloaded.

On the right, the meter can measure the voltage drop across a load, or can measure the voltage across any component, or group of components, that are a subset of the load.

A disadvantage shared by dedicated analog and digital volt meters is that they are not usually interchangeable between AC and DC.

How It Works

An analog volt meter usually contains a built-in, high-value fixed resistance, and an ammeter measuring the current passing through it. The current sensed by the ammeter is then converted to a reading in volts.

Suppose that R represents the fixed resistance, U is the voltage drop across the resistance, and I is the current flowing through it. Ohm's Law tells us:

```
U = I * R
```

The formula shows that if the resistance is fixed, the voltage will vary in proportion with the current, and therefore can be calculated from it.

Load-Related Inaccuracy

When the meter is measuring the voltage drop created by a load that has high resistance comparable to the internal resistance of the meter itself, the meter will not give an accurate reading. Figure 31-5 illustrates this problem.

On the left side of the figure, two resistors 10M each are wired in series between a 9VDC power source and negative ground. Because the resistors are equal, each of them imposes an equal voltage drop of 4.5V.

On the right side of the figure, an additional 10M resistance has been added in parallel with the lower resistor. When two resistors of values R1 and R2 are wired in parallel, their total resistance, R, is given by this formula:

```
1 / R = ( 1 / R1 ) + ( 1 / R2 )
```

Therefore, the total resistance in the bottom half of the circuit is now 5M instead of 10M, and the voltage drop across the upper resistor becomes twice the voltage drop across the two lower resistors.

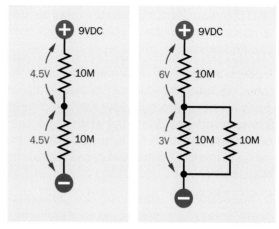

Figure 31-5 *If a meter measuring the voltage drop across a load has an internal resistance comparable to that of a load, the meter will not give an accurate reading. See text for details.*

Now suppose that the additional 10M resistor is actually the fixed resistance inside a volt meter. In fact, many volt meters do have an internal resistance of around 10M. Because the meter has reduced the resistance in the bottom half of the circuit by a factor of 2, it measures the voltage drop as 3V. If the meter had an ideal, infinite resistance, it would give the correct value of 4.5V. In the real world, that is impossible.

Bar Graph

Sometimes it is useful to represent voltage with a graphical display. A *bar graph* component makes this possible, and is often used in audio equipment.

The bar graph can consist of a row of *LEDs*. In a component designed for this purpose, there are usually 10 or more. To represent 0 volts, all the LEDs remain dark. More LEDs are illuminated as the voltage increases.

Because the bar graph itself only contains LEDs, a driver is necessary to convert voltage into the "rising thermometer" effect. Examples are the LM3914 (linear), the LM3915 (logarithmic, 3 dB per step), and the LM3916 (using VU or Volume Units, for audio). Each of these components has

10 LED outputs, and is fabricated using a chain of resistors as a multitap voltage divider, with 10 comparators.

A driver can be set to show only one LED corresponding with the current voltage level, or (more commonly) a cumulative string of LEDs from zero upward. Some bar graphs also contain different colored LEDs, such as green for the first 7, yellow for the next 2, and red for the final one.

A microcontroller containing an analog-to-digital converter can be used to illuminate a bar graph, instead of a driver chip.

An example of a bar graph is the Avago HDSP-4830, shown in Figure 31-6.

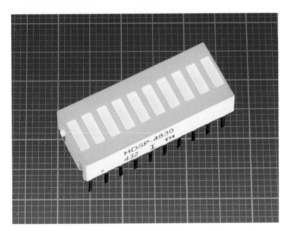

Figure 31-6 *A bar graph LED display, which can be used to represent voltage. The background grid is in millimeters.*

What Can Go Wrong

Confusing AC with DC

A panel meter that is designed only to measure DC should not be used with AC, and vice-versa. Erroneous readings or damage to the meter may result.

High Circuit Impedance

When a meter is measuring the voltage drop across a high-value resistance, it will give an inaccurate reading, as it is diverting a significant proportion of the current.

Voltage Out of Range

Attempting to measure a voltage that exceeds the limited range of a meter can damage the meter or blow its internal fuse, if it has one. Attempting to measure a voltage that is at the very low end of a meter's range will result in an inaccurate value.

Voltage Relative to Ground

If one input to a digital meter is connected with ground, the meter will only be able to measure voltage in a circuit relative to ground.

Sensor Output

A

This appendix provides some basic information about nine forms of sensor outputs, and the ways in which they can be processed. Other types of encoded output exist, but those examined here are the ones most likely to be found.

Figure A-1 provides an overview. Every sensor initially creates an analog output, which is sometimes connected directly to an output pin. In thermistors and photoresistors, for example, the internal resistance of the component constitutes its output. In many sensors, however, the behavior of the sensing element is processed internally to create voltage, open collector, encoded pulse stream, or current output.

If the sensor is chip-based, it may process the analog sensor response internally to create a binary output or a digital output.

In this Encyclopedia, the term "binary output" means "an output that has two states," usually logic-low or logic-high. The states may be accessible via an output pin, or may be processed internally to create a pulse stream. In that case the stream fluctuates between the two states as a way of encoding an analog value with pulse-width modulation (PWM) or frequency. Other types of encoding are also possible, but are unusual.

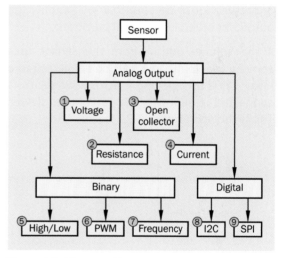

Figure A-1 *Nine possible types of output from sensors. The primary categories are highlighted in green.*

The term *digital output* is used here to mean one or two bytes of data that are stored in a register (memory location) in the sensor chip. While other forms of output are "always on" and can be accessed at any time from an output pin, a digital output is usually not available until an external device, such as a microcontroller, sends an instruction to the sensor chip, telling it to return the data. This two-way communication usually is handled by the I2C or SPI protocols (other protocols exist, but are less common).

Analog Outputs

Voltage is by far the most common form of analog output. Other forms of analog output can easily be converted to a voltage value, using the simple techniques described here.

1. Analog: Voltage

Direct Connection: Analog-to-Analog

An analog voltage output can be connected directly to an analog input, so long as the range is compatible and the sensor can provide sufficient current. Examples of external analog devices would be an analog volt meter, a light source or sound source that changes intensity, or a transistor or op-amp that will amplify the output for other audio/visual purposes.

If the voltage output from the sensor rises above a usable range, it can be converted to a lower value by applying it across two resistors connected in series to form a voltage divider. This is shown in Figure A-2.

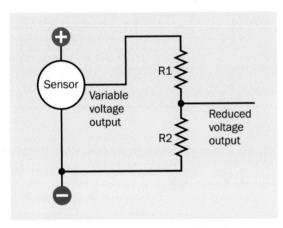

Figure A-2 *Using a voltage divider to reduce the range from a sensor.*

The values of R1 and R2 can be derived from this basic formula, where V_{SEN} is the voltage from the sensor and V_{OUT} is the output from the voltage divider:

Example A-1

$$V_{OUT} = V_{SEN} * (R2 / (R1 + R2))$$

The impedance of the device used for voltage sensing must be high relative to the values of R1 and R2. Note also that if the analog voltage out from the sensor varies linearly with the phenomenon being sensed, this relationship is likely to be disturbed by the voltage divider.

Analog-to-Binary Conversion

The term "binary" is used here to mean an output that can be in one of two states, such as logic-high and logic-low.

A varying analog voltage output can be simplified by passing it through a component that transforms the signal into binary form. This may be done by using a logic chip with a *Schmitt trigger* input, a *zener diode*, or a **comparator**. (For a description of comparators, see the entry in Volume 2.) A comparator provides desirable features such as adjustable positive feedback to create hysteresis. It may be used, for example, to convert a slowly changing signal from a phototransistor while the sun is setting, to a high/low output that can activate a relay to switch on a light.

Analog-to-Digital Conversion

The analog voltage output from a sensor can be digitized by an external analog-to-digital converter (ADC), either inside a microcontroller or by a separate ADC chip.

If a microcontroller is used, a sensor may often be wired directly to an input pin that connects internally with an ADC. A program in the microcontroller can then assess the integer output from the ADC and either execute a conditional statement or convert the value to a format appropriate for another device, such as a digital display.

If an ADC chip is used, there are thousands to choose from. A few basics:

- A *flash* converter contains a row of comparators with different reference voltages generated with a chain of equal resistors. The comparator outputs are fed into a priority **encoder** that outputs a binary number. This system is very fast but has limited resolution.

- A *successive approximation* converter uses a single comparator, comparing the input voltage with the output from a DAC. The binary number that is supplied to the DAC is determined one bit at a time, from the most significant to the least significant bit, using the comparator's result to determine if the bit should be 0 or 1. These bits are stored in a register, called a successive approximation register (SAR). When the process finishes, the SAR contains a binary representation of the input voltage. This type of ADC can achieve high resolution (many bits) at the cost of lower conversion speed.

- In a *dual slope* converter, a capacitor is charged for a fixed time, with a rate proportional to the input voltage, then discharged at a known rate while measuring time by counting clock pulses. The resulting count is the ADC output.

The name of the converter is derived from the voltage on the capacitor.

- A *voltage-to-frequency* converter uses a voltage-controlled oscillator to produce pulses with a frequency proportional to the input voltage. If the pulses are counted over a fixed time interval, the count is proportional to the signal level.

The number of the bits in the output from an ADC must be sufficient to digitize the input voltage range with desired accuracy. Because the voltage range may contain unexpected peaks, a cautious strategy is to use many more bits than are necessary. However, this means that for most of the time, only a few bits will be used to represent the low end of the voltage range, and accuracy will suffer.

For example, suppose a voltage input normally ranges from 0V to 2V, with occasional brief excursions to 8V. An 8-bit ADC can provide 256 digital values to represent input voltages. If the values are spread uniformly over the full 8V input range, the least significant bit can measure 1/32nd of a volt, or about 31mV. Smaller voltage fluctuations will be ignored. On the other hand, if the 256 values are used to measure a range of just 2 volts, the least significant bit can measure 1/128th of a volt, or slightly less than 8mV—but voltages higher than 2V will be clipped.

An ADC will typically require a reference voltage, and will digitize the range from 0V to that voltage. The reference voltage must be chosen with the issues of accuracy and range in mind.

A microcontroller may provide a feature in its language to perform automatic scaling of an analog input within limits set by a variable in program code. This is done by comparing the input with a selectable voltage level, such as the voltage of the power supply, an externally supplied voltage, or a fixed built in reference. While the ADC in the microcontroller normally digitizes values from 0V to the supply voltage for the chip, the conversion routine can instruct the microcontroller to use its full number of bits (often, 10) to digitize an input range from 0V to 1V.

For a higher sample rate, an ADC chip may be connected with the microcontroller over an I2C or SPI bus.

2. Analog: Resistance

Resistance-to-Voltage Conversion

A sensor that changes its resistance as it responds to its environment can be placed in a *voltage divider* to provide an analog voltage output. This is illustrated in Figure A-3.

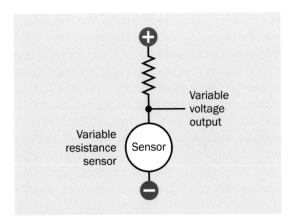

Figure A-3 *The basic principle of putting a variable-resistance sensor in series with a fixed resistance, to create a voltage divider.*

To choose the value for the series resistor, if R_{MIN} and R_{MAX} are the minimum and maximum resistance values for the sensor, the optimum value R_S for the series resistor, to produce the widest variation in voltages, will be found from this formula:

$$R_s = \sqrt{(R_{min} * R_{max})}$$

When the sensor has been set up in this way, the output can now be processed in the same way as any analog voltage output from a sensor.

The datasheet of the sensor should be consulted to make sure that a series resistor value does not allow any chance of the sensor being damaged by excessive current.

3. Analog: Open Collector

Many sensor packages or modules include a bipolar transistor that has an *open-collector* output (or *open drain*, if a CMOS transistor is used). The transistor may or may not be incorporated in an internal **op-amp** (described in Volume 2). Either way, the principle is the same.

Figure A-4 shows a sensor, in darker blue, that has one connection for positive power, another connection for negative ground, and a third connection to the collector of the internal transistor.

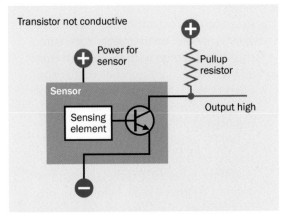

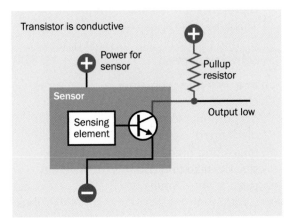

Figure A-4 *How to use an open-collector output from an internal transistor.*

In the upper section of the figure, the sensing element is not applying voltage to the base of the internal transistor, and the transistor only conducts a tiny amount of leakage current. Power applied to it through an external *pullup resistor* cannot reach negative ground in any significant quantity, and therefore it can provide a voltage input to a high-impedance device such as a microcontroller, or can power a component such as an LED, which draws relatively little current (20mA or less).

In the lower section of the figure, the sensing element is now applying voltage to the base of the transistor, drastically lowering its effective resistance. The transistor diverts current from the pullup resistor to ground, and the output appears to go low.

The type of sensing element will determine whether the transistor becomes conductive or nonconductive when the element detects a stimulus.

The value that should be chosen for the pullup resistor will depend on the impedance of any device attached to the open-collector output. A 10K resistor may be appropriate for use with a device such as a microcontroller, which has very high impedance. At the other extreme, if the device is an LED, a 330-ohm resistor may be necessary. The value of the pullup resistor must be sufficiently low to enable reliable operation, but sufficiently high to prevent excessive current from passing through the internal transistor when it becomes conductive (20mA is a common maximum value).

The voltage from the open collector can be processed in the same way as any analog voltage output from a sensor.

An open-collector output may be used when the outputs from multiple devices share the same bus. One device can drive the bus without the problem of other devices attempting to hold the bus voltage high.

4. Analog: Current

Relatively few sensors provide an output consisting of variations in current. Some semiconductor temperature sensors function in this way. The output current can be converted to a voltage output simply by placing a fixed series resistor, as shown in Figure A-5.

The voltage at the point shown, relative to negative ground, will vary linearly with the current. The value of the resistor should be defined in a datasheet for the sensor.

The voltage can now be processed in the same way as any analog voltage output from a sensor.

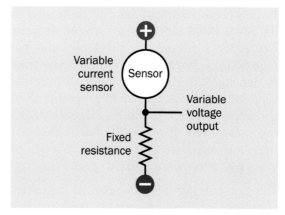

Figure A-5 *How to convert the output from a variable-current sensor.*

5. Binary: High/Low

A sensor that provides a binary output (that is, an output that is either logic-high or logic-low) can be connected directly with a microcontroller, if the voltage range is compatible. Program code in the microcontroller can then test the pin to establish its state. Note that some microcontrollers require a 3.3VDC power supply, while sensor chips may use 5VDC.

A binary output can also be used to control a solid-state relay, or an electromagnetic relay if transistor amplification is used. The output may be sufficient to power an LED indicator.

6. Binary: PWM

PWM is an abbreviation for *pulse-width modulation*. The sensor emits a stream of square-wave pulses with a fixed frequency, but the width of each pulse varies with the stimulus to which the sensor is responding. The width of each high pulse, relative to the wavelength between the start of one pulse and the start of the next, is called the *duty cycle*. A duty cycle of 0% means that there are no pulses at all. With a duty cycle of 100%, there are no gaps between pulses, so the output is high all the time. With a duty cycle of 50%, the duration of each high pulse is the same as the duration of the gap between pulses.

Various microcontrollers offer different ways to decode a PWM pulse stream. The most basic way is for a program to check an input pin repeatedly, as fast as possible, until a high state is detected. The microcontroller copies the value of its internal clock into a variable, then continues to check the input repeatedly until the pulse ends. The pulse duration that has been measured can be converted to a sensor value with a formula or lookup table.

This system is not recommended, as the microcontroller may miss the next pulse while it is converting the value of the previous pulse. To address this problem, a microcontroller language may offer a function that blocks execution of code while waiting for a pulse. The pulseIn() function on the Arduino is an example of this feature. However, the microprocessor must now spend most of its time waiting for a pulse instead of doing useful work.

A better solution is to write a program that is interrupt-driven.

Another option for decoding PWM is to use a low-pass filter that converts the pulse stream into an analog voltage, although some ripples will tend to remain.

Finally PWM can be used directly to power an LED or a DC motor, with transistor amplification as required. The speed of the motor or the brightness of the LED will vary with the duty cycle.

7. Binary: Frequency

Here again, the pulseIn() function on the Arduino may be used, so long as the frequency is a square wave with a known duty cycle.

8. Digital: I2C

In digital electronics, a *bus* is a communal pathway for sharing data among components or devices. The *I2C bus* is an abbreviation for *inter-integrated circuit* bus, developed originally by Philips in 1982. (Philips has since been subsumed into NXP Semiconductors.) The correct notation for I2C is I^{2}C, and it is spoken as "I-squared-C." However, the term is very commonly written as I2C.

The I2C standard defines a data sharing protocol that is limited to 400kHz (with some exceptions) and designed to work on a small scale, almost always within one device, and usually on one circuit board. It is a low-cost, simple design. Data is transmitted serially over two wires, and the devices sharing the bus are connected in parallel.

Typically there is one *master* device on the bus, and a number of *slave* devices. Masters and slaves can both transmit information, but the master normally initiates communication. It also emits a clock signal for synchronization of data.

A sensor is a slave device that can be interrogated by a microcontroller in its role as the master device. Because multiple slave devices can share a bus, the microcontroller needs a way to identify the slave that it wishes to talk to, and each slave is assigned a unique address for this purpose. Often a slave will allow the user to modify the last two bits of the address, so that up to four identical devices can share a bus.

Code libraries to support the I2C protocol are available for most microcontrollers, and communication with a sensor that uses I2C should simply require knowledge of the sensor's I2C address. However, the data registers in a sensor can be quite elaborate, requiring careful study of the manufacturer's datasheet. Multiple procedures may be required to set functions on a device (such as the sensitivity range of an accelerometer, or the threshold for a temperature alarm). Multiple procedures may also be required to read data out of a sensor (such as two bytes to define a temperature, and several bytes to obtain time as well as location readings from a GPS module).

9. Digital: SPI

SPI is an acronym for serial peripherals interface, a standard introduced by Motorola that serves a similar function to the I2C bus, described immediately above. The SPI standard is slightly more sophisticated, enabling duplex communication and higher data transfer speeds. However, SPI requires a minimum of three wires shared by all devices on the bus, and an additional device-selection line for each slave device. The benefit of the extra device lines is that devices are easier to select and address than on an I2C bus, where more program statements are required. As is the case for I2C, code libraries for microcontrollers are widely available to support SPI. However, the requirement for SPI to use three pins on a microcontroller, plus an additional pin for each slave device, is a disadvantage.

More sensors have I2C capability than SPI capability. The SPI protocol is potentially much faster than I2C.

A sensor that is SPI-enabled will almost certainly be available in a similar version that uses I2C. An increasing number of chip-based sensors support both protocols.

Glossary

This glossary is not comprehensive. It contains only the technical terms that have been used most frequently in this book in conjunction with sensor attributes.

ADC Analog-to-digital converter, which accepts a varying signal (usually, a voltage) as an input, and converts it to a digital value in the form of a binary number. This number is likely to have a high limit ranging from 255 decimal to 65535 decimal. Many microcontrollers contain their own ADC, which is multiplexed to assess the inputs on several pins. On an Arduino Uno, the ADC creates a digital value ranging from 0 through 1,023.

analog output When a sensor creates a voltage or varies its resistance without steps or increments, as a function of the phenomenon that it is measuring, this is an analog output.

binary output In this Encyclopedia, the term "binary output" describes a sensor output that only has two states: logic-high and logic-low, or on and off. The term is used in some datasheets, but more often a binary output is described, misleadingly, as an analog output.

breakout board A small printed-circuit board containing one or more integrated circuit chips, usually surface-mounted. The board makes the features of the chips easier to access, because it has pins or connectors with 2.54mm (0.1") spacing for convenient experimental use with a breadboard. Additional features may be included, such as a voltage regulator.

chip-based sensor This term is used in the Encyclopedia to describe a sensor that is etched into a silicon chip and usually has signal conditioning components and circuitry built in.

contact bounce The tiny and rapid vibrations of mechanical switch contacts, when the switch opens or closes. If the switch is connected with a digital device such as a logic chip, debouncing hardware may be necessary, to allow the contacts time to settle. If the switch is connected with a microcontroller, a delay from 5ms to 50ms may be written into program code. Different switches have widely different settling times.

decibel A unit that expresses relative power or intensity, often (but not exclusively) applied to audible sound. The decibel is one-tenth of a bel, and is abbreviated dB, with the B capitalized because it is derived from the name of Alexander Graham Bell. Because dB is a logarithmic unit, the scale has no zero origin. However, 0dB may be assigned arbitrarily to any intensity, in which case lower intensities will have a negative value. An increase of 3dB corresponds to a doubling of sound intensity (acoustic energy). However, when a sound is sensed by the human ear and evaluated by the brain, its subjective loudness doubles when the intensity increases by 10dB.

dielectric The insulating layer separating two plates in a capacitor.

hysteresis The difference between thresholds for switching an output on and off. When a sensor exhibits hysteresis, it may be unresponsive to a stimulus slightly above or below its equivalent current value. This may be useful to eliminate numerous responses to very small stimuli —for example, in a room thermostat.

I2C Interintegrated circuit bus. Sometimes written as I^{2}C, and often referred to verbally as "I-squared-C." A communications protocol that is often used between a microcontroller and other components on a circuit board. For a description, see "8. Digital: I2C" in the Appendix.

IMU Inertial measurement unit, consisting of three accelerometers and three gyroscopes, sometimes with the addition of three magnetometers. It can be used as a navigational aid. It may also be used in handheld user input devices, such as game controllers.

Kelvin A temperature scale, often abbreviated with the letter K, in which each degree is equivalent to a Celsius degree, but 0 degrees is at absolute zero—the temperature at which materials have no heat energy at all. 273 degrees K is approximately equal to 0 degrees C.

MEMS Microelectromechanical system, i.e., an integrated circuit chip that also contains tiny moving parts. For example, a MEMS accelerometer is built around microscopic springs that respond to accelerative forces.

newton A unit of force named after Isaac Newton, abbreviated with the capital letter N. A force of 1N will accelerate a mass of 1kg at a rate of 1 meter per second each second.

open collector output Many sensors have an open collector output, or contain an op-amp that has an open collector output. The output pin is attached to the collector of an internal transistor, with its emitter connected to negative ground. Positive voltage applied through a pullup resistor to the open collector will be grounded when the internal transistor is conducting current, but will be available for other devices when the transistor is off. See "3. Analog: Open Collector".

orthogonal Angled at 90 degrees. Three orthogonal elements in a sensor will all be angled at 90 degrees to each other.

pascal A unit of pressure equivalent to 1 newton of force per square meter.

PIR Passive infrared sensor. See Chapter 4.

pullup resistor A resistor that pulls up an output or input voltage in the absence of a signal. May be used in conjunction with an **open collector output**.

quadrature An encoding system for output from a pair of sensors. If the sensors are identified as A and B, four output combinations are possible: A high and B low; A high and B high; A low and B high; A low and B low. A common application is to show the direction of movement of a magnetic or optical pattern past the sensor pair.

reference temperature The temperature at which the output signal of a temperature sensor is measured. This is often listed in a datasheet.

register A section of memory that stores a digital value (usually 1 or 2 bytes in a sensor).

target Any object that is being detected by a motion sensor, proximity sensor, or presence sensor.

temperature coefficient The percentage increase or decrease in the value of a sensor as a result of unit change in temperature (usually 1 degree Celsius). Often abbreviated as TC. The value may be resistance, voltage, or current, depending on the sensor. The temperature coefficient should be negative if the value of the sensor diminishes when its temperature increases. If it is expressed in parts per million, abbreviated ppm, it can be converted to a percentage by dividing by 10,000.

Wheatstone bridge A network of four resistors. At least one of the resistors has an unknown value, while the others have precisely known reference values. The network enables calculation of the unknown value. See Figure 12-2.

Index

About the Authors

Charles Platt is the author of *Make: Electronics* and *Make: More Electronics*. He is a former senior writer for *Wired* magazine, and is a contributing editor to *Make:* magazine, for which he writes a column on electronics.

Fredrik Jansson is a physicist from Finland, with a PhD from Åbo Akademi University. He is currently living in The Netherlands, where he works on swarm robotics and simulates sea animals in the Computational Science group at the University of Amsterdam. Fredrik has always loved scavenging discarded household electronics for parts, and is a somewhat inactive radio amateur with the call sign OH1HSN. He also fact-checked Charles Platt's previous book, *Make: More Electronics*.

Colophon

The cover and body fonts are Myriad Pro, the heading font is Benton Sans, and the code font is Dalton Maag's Ubuntu Mono.